Excursions in Electromagnetics

A Study on the Formulation and the Reformulation of Maxwell's Equations

---

Krishna Srinivasan

First Edition

JOHANNES KEPLER vicariously continued the work of Ptolemy, as he tried to fit a circular orbit to Mars, on the data collected by Tycho Brahe. His result had an error of only eight arcminutes $8'$, equivalent to about a quarter of the width of the full moon in angular measure, as viewed from the earth. This "small" error, however, was not acceptable to Kepler, and led to the "reformation of all of astronomy". A famous quote from his book *Astronomia Nova* [1][2],

> "Since the divine benevolence has vouchsafed us Tycho Brahe, a most diligent observer, from whose observations the $8'$ error in this Ptolemaic computation is shown ... if I had thought I could ignore eight minutes ... I would already have made enough of a correction ... Now, because they could not have been ignored, these eight minutes alone will have led the way to the reformation of all of astronomy, and have constituted the material for a great part of the present work."

About this Book

My original goal was only to write an introductory book on electromagnetics. I was not quite expecting to be writing this book. There are only four equations to learn in electromagnetics. Yet, they seem like an "infinite reservoir of knowledge" that one could eternally draw from!

Although I wrote about Maxwell's equations in *An Electrifying Introduction to Electromagnetics*, the prerequisite for this book, I could not fully agree with the formulation of Faraday's law, until I could answer the following questions:

1. Why is Faraday's law formulated using the electric field $\vec{E}$, instead of the electric displacement (electric-flux density) $\vec{D}$?
2. What effect does a dielectric material have on the electric field generated from Faraday's law? How is this taken into account in Maxwell's equations?
3. What is the meaning of $\vec{D} = \epsilon_r \vec{E}$ (written in electrostatic units) in the context of Faraday's law?

Meditating countless hours on this problem, revealed deeper meanings on why Maxwell's equations are formulated the way they are, and the answers to the above questions. This has been my motivation to write this book.

An outcome of this research is a hypothesis, the reformulation of Faraday's law, of which, the present formulation is a special case. Modifying Maxwell's equations took many many trials to arrive at the formulation presented here.

Any proposal to reformulate Maxwell's equations, will surely be treated with a chuckle or disdain. I use this platform, therefore, to present my ideas, which I greatly value. This work is a reflection of my struggle to understanding Maxwell's equations, and looking beyond the equations! I dream of Maxwell-Srinivasan equations.

Krishna Srinivasan
2020

Contents

1 Evolution of Electromagnetic Units   **7**

2 Review of Maxwell's Equations in CGS Units   **9**
    2.1 Comparison of Units in ESU and EMU   10

3 Review of Maxwell's Equations in SI Units   **13**

4 Redefining Electric and Magnetic-Flux Densities $\vec{D}$ and $\vec{B}$, Without the Constants ϵ_o and μ_o   **15**

5 Introducing the Electric-Displacement Field (or Electric-flux Density) $\vec{D}$ **17**
    5.1 Gauss's Law in Free Space   17
    5.2 Faraday's Experimental Setup   18
    5.3 Experiment Results   19
    5.4 Explanation of Faraday's Experiment Results   19
    5.5 Introducing the Electric Displacement $\vec{D}$ 20

6 Polarization Charge and Polarization Vector   **22**
    6.1 Electric Dipole Moment   25
    6.2 Polarization Vector   25

7 Key Points on Electric Field $\vec{E}$ and Electric Displacement $\vec{D}$ **29**
    7.1 The Law of Conservation of Energy Equation in Electrostatics   30
    7.2 The units of $\vec{D}$ and $\vec{E}$ 30

8 Electric Displacement and Polarization Current in Ampere's Law   **31**

9 Introducing the Magnetic-Flux Density $\vec{B}$ **33**
    9.1 Ballistic-Galvanometer Experiment   33
    9.2 Magnetic-Flux Density $\vec{B}$ 34

10 Magnetization Vector $\vec{I}$ — 36
10.1 Magnetic Susceptibility — 36
10.2 The Definition of the Magnetic Dipole Moment With Magnetic Charges [Optional] — 37
10.3 The Magnetic Dipole Moment of a Current Loop — 38
10.4 Magnetization Vector $\vec{I}$ — 42

11 Key Points on Magnetic Field $\vec{H}$ and Magnetic-Flux Density $\vec{B}$ — 44

12 On the Formulation of Ampere's Law — 45

13 Rewriting Maxwell's Equations With Electric Field Separated by its Sources — 48
13.1 Sources of Electric Displacement $\vec{D}$ and Electric Field $\vec{E}$ — 48
13.2 Rewriting Maxwell's Equations — 48

14 Redefining $\vec{D}_C/\vec{D}_F$, $\vec{B}$, and $\vec{P}_C/\vec{P}_F$, Without the Constants ϵ_o and μ_o — 52

15 A New Perspective on the Meaning of Electric Displacement $\vec{D}_C$ — 55

16 On the Formulation of Faraday's Law — 57
16.1 Case (I): non-NULL effect — 57
16.2 Case (II): NULL effect — 58

17 The Reformulation of Faraday's Law Using Electric Displacement Instead of Electric Field — 59
17.1 Introducing a New Permittivity κ of the Faraday Electric Field — 59
17.2 A Thought Experiment Similar to Chapter 12 — 60

18 The Reformulated Maxwell's Equations — 64
18.1 The Law of Conservation of Energy — 65

19 Consequences of the Reformulation of Faraday's Law — 67
19.1 Characteristics Invariant to the Reformulation of Faraday's Law — 67
19.2 Changes Resulting From the Reformulation of Faraday's Law — 69
19.3 The Reformulated Faraday's Law and Magnetic Current Density — 69

20 The Wave Equation in the Reformulated Maxwell's Equations — 71

21 $\kappa \neq \epsilon_r$ — 73
21.1 A Note on Ampere's Law — 75
21.2 The Dependence of ϵ_r on the Electric-Field Pattern — 76

22 Measurement of ε_r and κ — 77
   22.1 Two Ways to Measure ϵ_r in the Existing Formulation of Maxwell's Equations — 77
   22.2 Measurement of ε_r and κ in the Reformulated Maxwell's Equations — 79
   22.3 ε_r and κ of Distilled Water — 80
   22.4 Experimental Verification of the Reformulated Maxwell's Equations — 81

23 The Derivation of the Lorentz Force Law from the Reformulated Faraday's Law — 82

24 Key Points on the Differences Between the Electric-Displacement Fields $\vec{D}_C$ and $\vec{D}_F$ — 84
   24.1 The Total Electric-Displacement Field $\vec{D}^*$ and $\vec{D}'$ — 85

25 A Comparison of the $\vec{B}/\vec{H}$ and the $\vec{D}/\vec{E}$ Relations — 87

26 Other Possibilities in the Reformulation of Maxwell's Equations — 89

27 Summary — 91

1
Evolution of Electromagnetic Units

Any physical quantity, such as length, velocity, time, etc. must be specified by a magnitude or the value of the quantity, a direction if its a vector, and the units associated with it. Without them, the value of the physical quantity is incomplete. For example, a stick of some length can serve as the definition of the unit length, with the new unit, say *stk*. The definition of the unit length as the length of the stick or its unit *stk*, both mean the same. The definition of the unit length, allows one to make a measurement of length. A measured value of length will be a multiple, and/or a fraction of this stick length, say 2 *stk* or 3.5 *stk*.

In the past, the CGS system of units was widely used. CGS stands for the initials of the 3 base units: centimeters (cm) for length, grams (gm) for mass, and seconds (s) for time. The units of other variables are made up of the base units, and are called derived units. The reader is referred to other books for more details on how the unit values of these base units are defined.

The *Système International d'Unités* (International System of Units), or SI units, is the system of units used today. In SI units, there are 7 base units: meter (m), kilogram (kg), second (s), ampere (A), kelvin, mole, candela. Meters, kilograms, and seconds are the counterparts of centimeters, grams, and seconds in the CGS system. In this book, the only base units used are meters, kilograms, seconds, and amperes. The reader is referred to other books for more details on the definitions of the kilogram, the meter, and the second. The ampere has been discussed in detail in Reference [3].

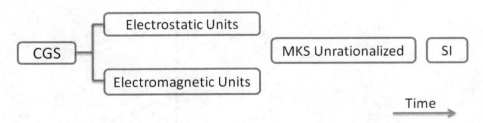

Figure 1.1: The evolution of electromagnetic units.

In electromagnetics, many different systems of units were used before the present SI units [4].

As shown in Figure 1.1, the electrostatic units (ESU) and the electromagnetic units (EMU) were widely used in the beginning. The units of the electrical quantities in ESU and EMU are made up of the CGS base units. The next major system used was the MKS unrationalized units, and finally our modern SI units. This has been explained in great detail in Reference [3], and will be summarized in Chapter 2 – Chapter 3. Learning how the units and the equations evolved over time is absolutely essential to understanding electromagnetics.

2
Review of Maxwell's Equations in CGS Units

This chapter is a review of the content in Reference [3]. Only the summary will be presented here. In ESU, the unit charge is defined first from the force between charges, or Coulomb's law, and the remaining electrical quantities are defined using the unit charge. In EMU, however, the current is first defined from the force between two parallel current-carrying wires, and the remaining electrical quantities are defined using the unit current.

ESU and EMU are not consistent with each other. The unit charge in ESU, for example, is not the same as the unit charge in EMU. The unit of charge in ESU and EMU, the statcoulomb and the abcoulomb, are also different.

The electromagnetic equations can be cast in a form that is common to both ESU and EMU, and using the constants ϵ_o and μ_o, which are the permittivity and the permeability of free space. Setting the constants ϵ_o and μ_o to the values shown in Table 2.1, would tailor the equations to

System of Units	ϵ_o	μ_o
ESU	1	$(3.0 \times 10^{10})^{-2}$ s^2/cm^2
EMU	$(3.0 \times 10^{10})^{-2}$ s^2/cm^2	1

Table 2.1: The values of ϵ_o and μ_o to tailor the common set of equations to ESU or EMU.

ESU or EMU. Note that the value of ϵ_o in ESU, and μ_o in EMU, are the dimensionless constant of 1. The value $(3.0 \times 10^{10})^{-2}$ s^2/cm^2, or the inverse speed of light squared, was first determined by Weber and Kohlrausch in 1856. Their experiment has been discussed in great detail in Reference [3].

The convention followed in this chapter is that the equations specific to ESU or EMU, will be marked beside the equation number; the equations common to both ESU and EMU will be marked as CGS; the equations common to both CGS and SI will be unmarked.

Maxwell's equations in the differential form is summarized in this chapter. Gauss's law satisfies the relation,
$$\nabla \cdot \vec{D} = 4\pi\rho, \tag{2.1, CGS}$$
where ρ is the volume-charge density, $\vec{D}$ is the electric-displacement field, also known as the electric-flux density. Ampere's law is
$$\nabla \times \vec{H} = 4\pi\vec{J} + \frac{\partial \vec{D}}{\partial t}, \tag{2.2, CGS}$$
where $\vec{J}$ is the volume current density, which is the current resulting from the flow of electric charges, and $\vec{H}$ is the magnetic field. Faraday's law is
$$\nabla \times \vec{E} = -\frac{\partial \vec{B}}{\partial t}, \tag{2.3}$$
where $\vec{E}$ is the electric field generated by the time-varying magnetic-flux density $\vec{B}$. The divergence free condition of $\vec{B}$ is
$$\nabla \cdot \vec{B} = 0. \tag{2.4}$$
The relation between magnetic-flux density $\vec{B}$ and magnetic field $\vec{H}$ is
$$\vec{B} = \mu_r \mu_o \vec{H}, \tag{2.5}$$
and μ_r is the relative permeability. The relation between $\vec{D}$ and $\vec{E}$ is
$$\vec{D} = \epsilon_r \epsilon_o \vec{E}, \tag{2.6}$$
where ϵ_r is the relative permittivity.

2.1 Comparison of Units in ESU and EMU

The units of the electrical quantities are different between ESU and EMU. In ESU, the charge is defined first from Coulomb's law,
$$\vec{F} = \frac{q_1 q_2}{r^2} \hat{r}, \tag{2.7, ESU}$$
where $\vec{F}$ is the force between two charges of magnitudes q_1 and q_2, separated by distance r, and $\hat{r}$ is the unit vector in the direction of the force. Applying dimensional analysis on the above equation, the unit of charge in ESU is
$$[q]_{ESU} = \frac{gm^{\frac{1}{2}} cm^{\frac{3}{2}}}{s}, \tag{2.8, ESU}$$
where the square brackets represents "the unit of". Since the current is the rate of flow of charges,
$$i = \frac{\Delta q}{\Delta t}, \tag{2.9}$$

where the charge Δq, flowing during the time Δt, results in the current i. Using dimensional analysis, the unit of current in ESU is

$$[i]_{ESU} = \frac{gm^{\frac{1}{2}} cm^{\frac{3}{2}}}{s^2}. \tag{2.10, ESU}$$

In EMU, since the current is first defined from the force between current-carrying wires,

$$\frac{F}{L} = \frac{2i_1 i_2}{d}, \tag{2.11, EMU}$$

where F is the magnitude of the force between two parallel wires of length L with negligible diameter of the wires, separated by distance d, carrying currents i_1 and i_2. Applying dimensional analysis on the above equation, the unit of current in EMU is

$$[i]_{EMU} = \frac{gm^{\frac{1}{2}} cm^{\frac{1}{2}}}{s}, \tag{2.12, EMU}$$

and the unit of charge in EMU, from Equation 2.9, is

$$\begin{aligned}[q]_{EMU} &= abampere \cdot s \\ &= gm^{\frac{1}{2}} cm^{\frac{1}{2}}.\end{aligned} \tag{2.13, EMU}$$

From the above units of charge and current, it can be seen that ESU and EMU don't have the same units for the electrical quantities. A summary of the units of some of the variables in ESU and EMU are shown in Table 2.2 and Table 2.3 [3] [4].

Quantity	Symbol	Units in ESU
Charge	q	$\frac{\text{gm}^{1/2}\text{cm}^{3/2}}{\text{s}}$
Current	i	$\frac{\text{gm}^{1/2}\text{cm}^{3/2}}{\text{s}^2}$
Magnetic-Flux Density	$\vec{B}$	$\frac{\text{gm}^{1/2}}{\text{cm}^{3/2}}$
Magnetic Field	$\vec{H}$	$\frac{\text{gm}^{1/2}\text{cm}^{1/2}}{\text{s}^2}$
Electric-Flux Density	$\vec{D}$	$\frac{\text{gm}^{1/2}}{\text{cm}^{1/2}\text{s}}$
Electric Field	$\vec{E}$	$\frac{\text{gm}^{1/2}}{\text{cm}^{1/2}\text{s}}$

Table 2.2: A summary of the units of some of the commonly used electrical quantities in ESU.

Quantity	Symbol	Units in EMU
Charge	q	$\text{gm}^{1/2}\text{cm}^{1/2}$
Current	i	$\frac{\text{gm}^{1/2}\text{cm}^{1/2}}{\text{s}}$
Magnetic-Flux Density	$\vec{B}$	$\frac{\text{gm}^{1/2}}{\text{cm}^{1/2}\text{s}}$
Magnetic Field	$\vec{H}$	$\frac{\text{gm}^{1/2}}{\text{cm}^{1/2}\text{s}}$
Electric-Flux Density	$\vec{D}$	$\frac{\text{gm}^{1/2}}{\text{cm}^{3/2}}$
Electric Field	$\vec{E}$	$\frac{\text{gm}^{1/2}\text{cm}^{1/2}}{\text{s}^2}$

Table 2.3: A summary of the units of some of the commonly used electrical quantities in EMU.

3
Review of Maxwell's Equations in SI Units

This chapter is a review of the contents in Reference [3]. In EMU, the resistance has the same unit as velocity! To avoid such meaningless units, the ampere is the base unit of current in SI. The units of other electrical quantities are derived units, made up of the base units, meter, kilogram, second, and ampere.

Maxwell's equations in SI are in the rationalized form, without the "spurious eruption of 4πs", in the words of Oliver Heaviside. Gauss's law, Faraday's law, the divergence-free condition of magnetic-flux density, and Ampere's law, in the differential form are

$$\nabla \cdot \vec{D} = \rho \tag{3.1}$$

$$\nabla \times \vec{E} = -\frac{\partial \vec{B}}{\partial t} \tag{3.2}$$

$$\nabla \cdot \vec{B} = 0 \tag{3.3}$$

$$\nabla \times \vec{H} = \vec{J} + \frac{\partial \vec{D}}{\partial t}. \tag{3.4}$$

The above equations are valid for both static and time-varying fields. The names of the field variables and their units are shown in Table 3.1. The relation between $\{\vec{D}, \vec{E}\}$ and $\{\vec{B}, \vec{H}\}$ are

$$\vec{D} = \epsilon_r \epsilon_o \vec{E} \tag{3.5}$$

$$\vec{B} = \mu_r \mu_o \vec{H}. \tag{3.6}$$

The values of ϵ_o and μ_o, the permittivity and the permeability of free space, are summarized in Table 3.2.

Electrical Quantity (Symbol)	Unit (Symbol)	Units
Current (i)	ampere (A)	A
Current density $(\vec{J})$	–	$A\,m^{-2}$
Charge (q)	coulomb (C)	$A\,s$
Volume charge density (ρ)	–	$C\,m^{-3}$
Magnetic-flux density $(\vec{B})$	tesla (T)	$N A^{-1} m^{-1}$
Magnetic field $(\vec{H})$	–	$A\,m^{-1}$
Electric field $(\vec{E})$	–	$N C^{-1}$
Electric-flux Density $(\vec{D})$	–	$C\,m^{-2}$

Table 3.1: The SI units of some of the electrical quantities.

System of Units	ϵ_o	μ_o
SI	$8.85 \times 10^{-12}\ C^2/(N \cdot m^2)$	$4\pi \times 10^{-7}\ N/A^2$

Table 3.2: The values of ϵ_o and μ_o in SI units.

4

Redefining Electric and Magnetic-Flux Densities $\vec{D}$ and $\vec{B}$, Without the Constants ϵ_o and μ_o

The exercise in this chapter is useful for the discussion in the future chapters. In ESU, repeating Equation 2.6, electric displacement $\vec{D}$ and electric field $\vec{E}$ are related as

$$\vec{D} = \epsilon_r \vec{E}, \tag{4.1, ESU}$$

since $\epsilon_o = 1$, noted in Table 2.1. ϵ_r has no units, and therefore, $[\vec{D}]$ and $[\vec{E}]$ have the same units, documented in Table 2.2. Likewise, in EMU, repeating Equation 2.5, magnetic-flux density $\vec{B}$ and magnetic field $\vec{H}$ are related as

$$\vec{B} = \mu_r \vec{H}, \tag{4.2, EMU}$$

since $\mu_o = 1$, noted in Table 2.1. μ_r has no units, and therefore, $[\vec{B}]$ and $[\vec{H}]$ have the same units, documented in Table 2.3.

In SI units, its possible to rewrite (not reformulate!) the equations, so that

$$[\vec{B}] = [\vec{H}], \tag{4.3}$$

and

$$[\vec{D}] = [\vec{E}], \tag{4.4}$$

have the same units. A simple way to rewrite this is to replace electric and magnetic-flux densities as

$$\vec{D} \to \epsilon_o \vec{D}^* \tag{4.5}$$
$$\vec{B} \to \mu_o \vec{B}^*, \tag{4.6}$$

where

$$\vec{D}^* = \epsilon_r \vec{E}, \tag{4.7}$$
$$\vec{B}^* = \mu_r \vec{H}. \tag{4.8}$$

The convention followed is that the asterisk superscript denotes the variable definition, without the constant ϵ_o or μ_o. Faraday's law, for example, can be written as

$$\nabla \times \vec{E} = \mu_o \frac{\partial \vec{B}^*}{\partial t}, \tag{4.9}$$

instead of Equation 3.2.

Rewriting the electromagnetic equations using $\vec{D}^*$ and $\vec{B}^*$, there is little change to the units of the electrical quantities. Walking through the SI definitions flowchart in Reference [3], or in Figure 19.1, and applying dimensional analysis, it can be verified that the units of variables in Table 3.1, except for $\vec{D}$ and $\vec{B}$, stay the same. From Equation 4.7,

$$[\vec{D}^*] = [\vec{E}], \tag{4.10}$$

and from Equation 4.8,

$$[\vec{B}^*] = [\vec{H}], \tag{4.11}$$

shown in Table 4.1.

Electrical Quantity (Symbol)	Unit (Symbol)	Units
Current (i)	ampere (A)	A
Current density ($\vec{J}$)	–	$A\,m^{-2}$
Charge (q)	coulomb (C)	$A\,s$
Volume charge density (ρ)	–	$C\,m^{-3}$
Magnetic-flux density ($\vec{B}^*$)	–	$A\,m^{-1}$
Magnetic field ($\vec{H}$)	–	$A\,m^{-1}$
Electric field ($\vec{E}$)	–	$N\,C^{-1}$
Electric-flux Density ($\vec{D}^*$)	–	$N\,C^{-1}$

Table 4.1: SI units of the electrical quantities in Maxwell's equations.

5

Introducing the Electric-Displacement Field (or Electric-flux Density) $\vec{D}$

The meaning of the electric field $\vec{E}$, is much simpler than the electric displacement $\vec{D}$, also known as the electric-flux density,

$$\vec{E} = \frac{\vec{F}}{q}, \tag{5.1}$$

where $\vec{F}$ is the force acting on an electric charge q. At any point, no matter what material is present, if the electric field is $\vec{E}$, a charge q experiences a force

$$\vec{F} = q\vec{E}. \tag{5.2}$$

Electric displacement is often introduced by jumping the gun to the definition of the polarization vector, and the bound charge. What often follows the barrage of definitions, is the "surprising" formulation of Gauss's law in a material medium. This flow makes it very difficult to follow along, without a logical development of electromagnetics.

Faraday's study of dielectric materials using spherical capacitors, led to the introduction of a new vector field, the electric displacement $\vec{D}$, and the relative permittivity ϵ_r of a material [3][5][6]. The contents of this chapter have been discussed in detail in Reference [3], and many of the details may not be included here to avoid repetition.

5.1 Gauss's Law in Free Space

Coulomb's law is used to define the unit electric charge from the equation,

$$\vec{F} = \frac{1}{4\pi\epsilon_o} \frac{q_1 q_2}{r^2} \hat{r}, \tag{5.3}$$

where $\vec{F}$ is the force between charges q_1 and q_2, separated by distance r, and $\hat{r}$ is the unit vector in the direction of the force. Coulomb's law will be limited to charges in free space. Using Coulomb's law and the definition of the electric field in Equation 5.1, Gauss's law for charges in free space in the electrostatic case, can be derived as

$$\oint_S \epsilon_o \vec{E} \cdot d\vec{A} = q_{enc}, \tag{5.4}$$

where S is a closed 3D surface enclosing charge q_{enc}. The derivation of the above equation in ESU is presented in Reference [3]. However, similar steps can be followed to derive the above equation in SI, and is not repeated here. Faraday's experiment, and the derivation of Gauss's law that is valid in any material medium, is presented next.

5.2 Faraday's Experimental Setup

The spherical capacitor used by Faraday is shown in Figure 5.1. It has an opening at the top by which dielectric materials, such as distilled water or melted wax, can be filled in the cavity between the spheres.

Figure 5.1: The spherical capacitor used by Faraday [5][6].

The experimental setup is shown in Figure 5.2. Faraday used two identical capacitors, A and B, with the two inner metal spheres, p and p', connected together by a wire, as well as the two outer metal spheres q and q'. A metal is an equipotential volume, and so are metal objects connected together. Therefore, p and p' are at the same potential, and so are q and q'. Since the inner spheres are connected, as well as the outer spheres, the potential difference between the inner sphere and the outer sphere in A, V_{pq}, is equal to that of B, $V_{p'q'}$,

$$V_{pq} = V_{p'q'}. \tag{5.5}$$

Since the capacitors are identical, and voltage is the path integral of the electric field, the electric fields in the cavities of A and B, $\vec{E}_A$ and $\vec{E}_B$, are equal,

$$\vec{E}_A = \vec{E}_B, \tag{5.6}$$

at the same point relative to the respective centers of A and B, in each of the cavities. This is true, independent of the dielectric material that fills the cavities of A and B.

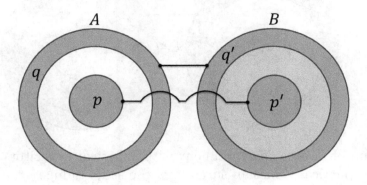

Figure 5.2: The experimental setup to study the variation of the charge stored in a capacitor with different dielectric materials.

5.3 Experiment Results

In the cavity of A, the material is always kept as air, but the material in B is varied. Capacitor A with the unfilled cavity acts as a reference for the experiment. The capacitors are charged simultaneously by connecting the inner and the outer spheres to the terminals of a Wimshurst machine. A detailed explanation of how a Wimshurst machine works is presented in Reference [3].

Faraday studied the ratio of the charge stored in A, Q_A, and B, Q_B. A way by which the quantity of charge stored in a capacitor can be measured, is using a ballistic galvanometer. The detailed methodology is presented in Reference [3]. If the dielectric material in both A and B is air, by symmetry, the charge stored in the two identical capacitors are equal. The ratio of the charge stored $Q_B : Q_A$ is 1.

However, in the case when B is filled with a dielectric material other than air, Faraday observed that the ratio is greater than 1. This ratio has a special name, and is called the relative permittivity of the dielectric material, denoted by the symbol ϵ_r, where r stands for *relative*, and it means the permittivity of a material relative to air. To be precise, the cavity in Capacitor A must be vacuum, which is the absence of any material, including air. The reference dielectric material of air will be assumed for simplicity.

5.4 Explanation of Faraday's Experiment Results

Faraday observed that Capacitor B, whose cavity is filled with a dielectric material, stores ϵ_r times more charge than Capacitor A, whose cavity is unfilled, or contains air. Faraday's experimental setup is shown in Figure 5.3. If $\vec{E}_A$ is the electric field in Capacitor A, applying Gauss's law in

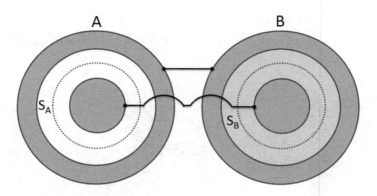

Figure 5.3: The setup of Faraday's experiment with spherical capacitors. S_A and S_B, shown by the dotted lines, are spherical Gaussian surfaces of the same radius.

free space, in Equation 5.4,

$$\oint_{S_A} \epsilon_o \vec{E}_A \cdot d\vec{A} = Q_A, \tag{5.7}$$

where $\pm Q_A$ is the charge stored in the spheres of A, which reside on the outside of the inner sphere, and the inside of the outer spherical shell [3], and S_A is the spherical Gaussian surface in the cavity, as shown by the dotted line. Solving the above equation, the electric-field strength in the cavity is $\propto Q$.

If this same equation is applied to Capacitor B, one could expect the field strength in the cavity $\epsilon_r \times$ greater, since $\epsilon_r \times$ more charge is stored in Capacitor B. By definition, voltage is the path integral of the electric field, and therefore, $V_{p'q'}$ is $\epsilon_r \times$ greater than V_{pq}. Since $V_{pq} = V_{p'q'}$, this contradicts the experiment results. This can be resolved with the explanation presented next.

If the dielectric material in the cavity of Capacitor B, reduces the electric-field strength by ϵ_r, then the electric fields in the cavities of A and B are equal. If the electric fields are equal, this means that the voltage $V_{pq} = V_{p'q'}$. This explains the reason that more charge is present in Capacitor B: the additional charge is present to overcome the reduction in the electric field caused by the dielectric material, so that the electric fields in the cavities of A and B are equal. This explanation can be captured as

$$\vec{E}_{material} = \frac{\vec{E}_{air}}{\epsilon_r}, \tag{5.8}$$

where $\vec{E}_{material}$ is the electric field at any point in a material with permittivity ϵ_r, and $\vec{E}_{air}$ is the electric field in air, or the electric field that would have existed at that point, if the material at that point does not reduce the field. This observation will be used to formulate Gauss's law in the next section, taking into account the reduction of electric-field strength in a dielectric material.

5.5 Introducing the Electric Displacement $\vec{D}$

The ratio of the charges stored in A and B is the relative permittivity ϵ_r,

$$Q_A = \frac{Q_B}{\epsilon_r}. \tag{5.9}$$

Since the potential difference between the outer and the inner conductors are the same in both the identical capacitors, as noted in Section 5.2 – Section 5.3,

$$\oint_{S_B} \epsilon_o \vec{E}_B \cdot d\vec{A} = \oint_{S_A} \epsilon_o \vec{E}_A \cdot d\vec{A}, \tag{5.10}$$

where S_B is a spherical Gaussian surface lying in the cavity of B, and of the same radius as S_A. From the above equations,

$$\oint_{S_B} \epsilon_o \vec{E}_B \cdot d\vec{A} = \frac{Q_B}{\epsilon_r}. \tag{5.11}$$

Rearranging the above equation,

$$\oint_{S_B} \epsilon_r \epsilon_o \vec{E}_B \cdot d\vec{A} = Q_B, \tag{5.12}$$

is the general form of Gauss's law.

In this example, the Gaussian surface is present in an uniform dielectric material. ϵ_r is moved inside the integral, which will account for the variation in the dielectric material over S. Gauss's law is also valid in this case. However, the validity of Gauss's law in any type of material medium, uniform or non-uniform, isotropic or anisotropic, linear or non-linear, or in the case of time-varying fields, can be proven with the current-continuity equation, and is presented in Reference [3] (See also [7][8]). In other words, Gauss's law is always valid!

The integrand in the above equation is assigned a new vector-field quantity $\vec{D}$, and is called the electric displacement,

$$\vec{D} = \epsilon_r \epsilon_o \vec{E}. \tag{5.13}$$

$\vec{D}$ and $\vec{E}$ are related at any point by the above equation. ϵ_r is a property of the material at that point, as seen in Faraday's experiments with spherical capacitors. From the above equations, the general form of Gauss's law is written as

$$\oint_S \vec{D} \cdot d\vec{A} = q_{enc}, \tag{5.14}$$

where S is the Gaussian surface enclosing charge q_{enc}. The conventional definition of electric displacement $\vec{D}$ is Equation 5.13, which must satisfy Gauss's law in Equation 5.14. New perspectives on the different meanings of electric displacement will be presented in Chapter 15 and Chapter 24.

6

Polarization Charge and Polarization Vector

A clear understanding of the necessity to introduce a new vector field, electric displacement $\vec{D}$, can be achieved by studying Faraday's experiment with spherical capacitors, and different dielectric materials, presented in Chapter 5. The experiment results show that the electric field is modified by the dielectric material, reducing its strength by ϵ_r, captured in Equation 5.8. The atomic view of a dielectric material can be used to explain this behavior, which leads to the polarization vector and the bound charge, explained next.

A simplistic view of the atoms in a dielectric material, provides an understanding of how a dielectric material reduces the strength of an electric field. In the presence of an electric field, the positive and the negative charges in the atoms become distorted, as illustrated in Figure 6.1. This polarization of the atoms of the dielectric material, has an effect on the electric field, modifying the field.

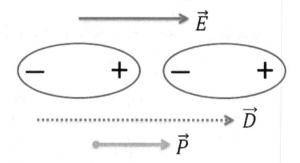

Figure 6.1: The reduction in the electric-field strength in a dielectric material, due to the polarization of the atoms.

Equation 5.13 relates $\vec{D}$ and $\vec{E}$ with a scaling factor ϵ_r. An alternate way is to relate $\vec{D}$ and $\vec{E}$ as a sum,

$$\vec{D} = \epsilon_o \vec{E} + \vec{P}, \tag{6.1}$$

where $\vec{P}$ is called the polarization vector. The units of the left-hand side and the right-hand side must match, and therefore, the unit of $\vec{P}$ must be the same as $\vec{D}$. The reason for formulating the equation in this manner, instead of writing the right-hand side as

$$\epsilon_o \vec{E} + \epsilon_o \vec{P}, \tag{6.2}$$

where ϵ_o is not included in the definition of $\vec{P}$, will be discussed at the end of this chapter.

If $\vec{P}$ is written as

$$\vec{P} = \epsilon_o \chi_e \vec{E}, \tag{6.3}$$

where χ_e is known as the electric susceptibility with no units, substituting the above equation in Equation 6.1,

$$\vec{D} = \epsilon_o \left(1 + \chi_e\right) \vec{E}. \tag{6.4}$$

Equating the above equation to Equation 5.13,

$$\vec{D} = \epsilon_o \epsilon_r \vec{E}, \tag{6.5}$$

the relative permittivity can be related to electric susceptibility as

$$\epsilon_r = 1 + \chi_e. \tag{6.6}$$

Since $\epsilon_r \geq 1$, χ_e is a positive number. Calculating the flux across a closed 3D surface S on both sides of Equation 6.1,

$$\oint_S \vec{D} \cdot d\vec{A} = \oint_S \epsilon_o \vec{E} \cdot d\vec{A} + \oint_S \vec{P} \cdot d\vec{A}. \tag{6.7}$$

Substituting Gauss's law in the left-hand side of the above equation,

$$q_{enc} = \oint_S \epsilon_o \vec{E} \cdot d\vec{A} + \oint_S \vec{P} \cdot d\vec{A}, \tag{6.8}$$

where q_{enc} is the charge enclosed by S. Rearranging the above equation,

$$\oint_S \epsilon_o \vec{E} \cdot d\vec{A} = q_{enc} - \oint_S \vec{P} \cdot d\vec{A}. \tag{6.9}$$

From dimensional analysis, the flux across the surface S due to the polarization vector $\vec{P}$, must have the unit of charge $[q]$, denoted as,

$$\left[\oint_S \vec{P} \cdot d\vec{A}\right] = [q], \tag{6.10}$$

since it is subtracted from q_{enc}. Equation 6.9 is similar to Gauss's law, except that the left-hand side is written as the flux due to $\vec{E}$, rather than $\vec{D}$. The enclosed charge by the closed surface in the equation is the total charge, including both the free charge and the bound charge,

$$\oint_S \epsilon_o \vec{E} \cdot d\vec{A} = q_{enc} + q_b, \tag{6.11}$$

unlike Gauss's law, where the enclosed charge is only the free charge. q_b is known as the polarization charge, or the *bound* charge,

$$q_b = -\oint_S \vec{P} \cdot d\vec{A}. \tag{6.12}$$

The bound charge is viewed as the charge in one of the two charge polarities, in the atomic dipoles making up the dielectric material, lying within the surface S.

The reason for the name *bound* charge makes sense, since the charge in the atomic dipole of the dielectric material is not free to flow. On the other hand, q_{enc} is known as free charge, since this charge is free to flow from one conductor to another. q_{enc} is sometimes denoted as q_f for *free* charge.

The bound charge is illustrated in Figure 6.2, using the spherical capacitor example. The cross section of a spherical capacitor is shown in the figure, whose cavity is filled with a dielectric material of relative permittivity ϵ_r. For example, the inner sphere is charged positive, and the outer is charged negative. As explained in Reference [3], the positive charges lie on the outer surface of the inner sphere, and the equal and opposite negative charges, on the inner surface of the outer shell, as marked.

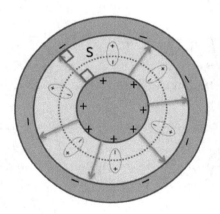

Figure 6.2: The cross section of a charged spherical capacitor, and the polarized atoms of the dielectric material.

The electric-field lines from the inner sphere to the outer shell in the cavity, is shown by the arrows. Since χ_e is a positive number, from Equation 6.3, $\vec{P}$ is in the same direction as $\vec{E}$. The spherical surface S in Equation 6.9, is marked by the dotted line. By spherical symmetry, the magnitude of the polarization vector and electric field are constants on the surface S, and are perpendicular to the spherical surface. Substituting Equation 6.3 in Equation 6.12, the resulting value is

$$q_b = -4\pi\epsilon_o \chi_e r^2 E, \tag{6.13}$$

where E is the magnitude of the electric field, on the spherical surface S of radius r. Note that q_b is a negative value, consistent with the negative charge of the dipoles within the surface S, in

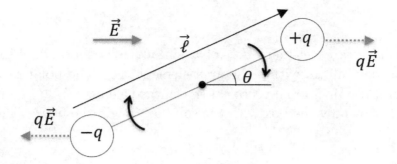

Figure 6.3: An electric dipole made up of charges $\pm q$.

the figure, for the direction of the electric field assumed in this example. Using Gauss's law, it is left as an exercise for the reader to verify that

$$q_f = 4\pi\epsilon_o\epsilon_r r^2 E, \tag{6.14}$$

where q_f is the free charge on the inner sphere, and is a positive value for the assumptions made. From the above equations, the ratio of the bound charge to the free charge is

$$\frac{q_b}{q_f} = -\frac{\chi_e}{\epsilon_r}. \tag{6.15}$$

6.1 Electric Dipole Moment

An electric dipole is made up of two tiny charged metal spheres of electric charges $\pm q$, and q is positive, joined together by a dielectric rod, shown in Figure 6.3. The dielectric rod is free to rotate on the plane of the page about its center pivot. $\vec{\ell}$ is the length vector from $-q$ to $+q$. In an electric field $\vec{E}$, the force on the charges $\pm q\vec{E}$ creates a torque,

$$\vec{\tau} = q\vec{\ell} \times \vec{E}, \tag{6.16}$$

where

$$\vec{p} = q\vec{\ell} \tag{6.17}$$

is known as the electric dipole moment. Applying dimensional analysis on Equation 6.17, the unit of the electric dipole moment in SI units is

$$[\vec{p}] = C\,m. \tag{6.18}$$

6.2 Polarization Vector

Analyzing Equation 6.9, results in a different interpretation of the polarization vector. As shown in Figure 6.1, $\vec{P_i}$ will be related to the dipoles formed in a dielectric material, by the presence of an electric field.

An infinitesimal surface-area element dA_i of a surface S is shown in Figure 6.4(a). $\{\vec{D}_i, \vec{E}_i, \vec{P}_i\}$ are assumed to be uniform within the area element dA_i. The polarization vector $\vec{P}_i$ is at an angle θ_i to the outward unit-normal vector of the surface element $\hat{n}_i$. $\vec{P}_i$ and the surface dA_i form a parallelepiped, as shown by the box with the solid outline in the figure. The flux across dA_i due to $\vec{P}_i$,

$$\vec{P}_i \cdot d\vec{A}_i = P_i \, dA_i \cos\theta_i, \tag{6.19}$$

is also the volume of the parallelepiped. The volume has the unit of charge, as noted in Equation 6.10. The parallelepiped can be viewed as a blob of bound charge marked $+dq_i$,

$$dq_i = P_i \, dA_i \cos\theta_i. \tag{6.20}$$

The bound charge, however, as defined in Equation 6.12, is calculated using $-\vec{P}_i$, which is the opposite direction to the vector marked in the figure. A mirror image of the blob of charge, shown by the dotted parallelepiped lying *within* the surface, which represents $-dq_i$ is drawn. This blob of charge $-dq_i$, reduces the enclosed charge within the surface S, q_{enc}, to satisfy the flux across the surface caused by $\vec{E}$ in Equation 6.9.

The bound charge $\pm dq_i$ viewed as an electric dipole is shown in Figure 6.4(b). The electric dipoles model the charge separation in the atoms of the dielectric material, as noted in Figure 6.1. The positive and negative charges of an electric dipole are distinguished by circles and diamonds. $\vec{\ell}_i$ is a vector in the same direction as $\vec{P}_i$, and an arbitrary magnitude, straddling across the boundary of the surface, and separating the dipole charges. Multiplying both sides of Equation 6.20 by length ℓ_i, the magnitude of $\vec{\ell}_i$, the magnitude of the electric dipole moment is

$$dq_i \, \ell_i = \vec{P}_i \cdot d\vec{A}_i \, \ell_i \tag{6.21}$$
$$= P_i \, dA_i \cos\theta_i \, \ell_i, \tag{6.22}$$

where P_i and dA_i are the magnitudes of the vectors $\vec{P}_i$ and $d\vec{A}_i$. But

$$\ell_i \, dA_i \cos\theta_i = dV_i, \tag{6.23}$$

the volume of the parallelepiped formed by ℓ_i and dA_i, containing the dipole charge $\pm dq_i$. Rewriting the above equation,

$$P_i = \frac{dq_i \, \ell_i}{dV_i}. \tag{6.24}$$

Since the direction of $\vec{P}_i$ is the same as $\vec{\ell}_i$, the above equation can be written as a vector,

$$\vec{P}_i = \frac{dq_i \, \vec{\ell}_i}{dV_i}, \tag{6.25}$$

where $dq_i \, \vec{\ell}_i$ is the electric dipole moment, and the right-hand side is the electric dipole moment per unit volume. Note that the value of the right-hand side does not vary with ℓ_i, since ℓ_i in the numerator, cancels the ℓ_i in the denominator in Equation 6.23. From the above equation, the polarization vector $\vec{P}_i$ can be viewed as an electric dipole moment per unit volume.

A volume of dielectric material is discretized into volume elements, shown in Figure 6.5. Each of the volume elements contains many many atoms, and each of the atoms forming a dipole with an applied electric field. Lets say there are N of them, each with a dipole moment $\vec{p}_i$. $\vec{P}_i$ can be viewed as the average of all the dipole moments in a volume element, with volume ΔV,

$$\vec{P}_i = \frac{\frac{1}{N}\sum_{i=1}^{N}\vec{p}_i}{\Delta V}. \tag{6.26}$$

Applying dimensional analysis on the above equation, and from Equation 6.18, the unit of $[\vec{P}]$ is

$$[\vec{P}] = \frac{C}{m^2}, \tag{6.27}$$

which is the same as $[\vec{D}]$ in Table 3.1, as noted earlier. From Table 3.2, ϵ_o is a constant with units. If ϵ_o is not included as part of the definition of $\vec{P}$ in Equation 6.1, the polarization vector $\vec{P}$ cannot be viewed as the electric dipole moment per unit volume, since the dimensions would be incorrect. This is the reason for writing Equation 6.1 in this manner, and not defining $\vec{P}$ as Equation 6.2.

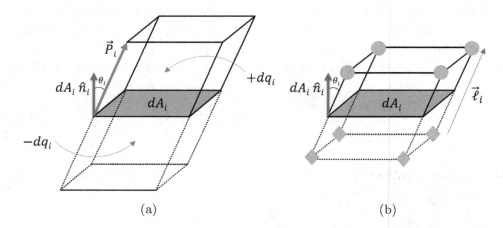

Figure 6.4: (a) The polarization vector at a surface element dA_i. (b) The flux due to the polarization vector across dA_i, viewed as an electric dipole moment.

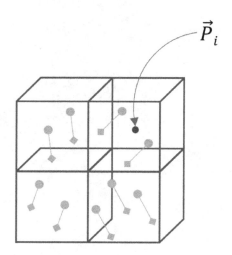

Figure 6.5: The polarization vector viewed as the electric dipole moment per unit volume.

7

Key Points on
Electric Field $\vec{E}$ and Electric Displacement $\vec{D}$

The conclusion that can be drawn from Faraday's experiments with spherical capacitors, is that the dielectric material at any point, has an effect on the electric field *at that point*, modifying the field. This is captured in Equation 5.8, where the dielectric material in the capacitor, reduces the strength of the electric field by ϵ_r.

If an electric charge is placed at any point, it experiences a force, whose value is captured in Equation 5.2,
$$\vec{F} = q\vec{E}. \tag{7.1}$$
The above equation always holds true, at any point, and at any time instant, whether there is a static or a time-varying field, linear or a non-linear material, etc. Therefore, what is implicitly stated is that, by definition, the electric field $\vec{E}$ at any point, includes the effect of the dielectric material at that point. If this is not the case, Equation 7.1 no longer holds true. This observation is very important, and will be revisited in Chapter 17.

Maxwell's equations, written as a set of differential equations, is valid at any point. The electric-field $\vec{E}$ in a differential equation, includes the effect of the dielectric material at that point. The effect of a dielectric material in modifying the electric field is assumed to be a localized effect, where the dielectric material *at a point*, modifies the electric field *at that point*. From the simplified atomic perspective of a dielectric material in Chapter 6, this can be justified. The polarization of atoms with no net charge, in a tiny localized region, modifies the electric field in that region.

The dielectric materials in the entire problem region, is taken into account in the electric-field solution, when the system of equations are solved together in the spatial region. Although numerical methods are not covered in this book, the spatial region is discretized into small mesh elements. The system of equations corresponding to the mesh elements are solved together, with the appropriate boundary conditions, and sources, thereby accounting for all the dielectric materials in the entire region, in the electric field $\vec{E}$ solution.

7.1 The Law of Conservation of Energy Equation in Electrostatics

Since $\vec{E}$ is the net electric field, including the effect of the dielectric materials, it was proven in electrostatics in Reference [3], the path integral over a loop ℓ,

$$\oint_\ell \vec{E} \cdot d\vec{l} = 0, \tag{7.2}$$

must be met, to not violate the law of conservation of energy. The above equation does not apply to $\vec{D}^*$, for example, since $\vec{D}^*$ is not the net electric field.

The above integral equation can be converted to the differential form. Using Kelvin-Stokes theorem, the path integral over the loop ℓ, can be written as an area integral over the surface S enclosing the loop,

$$\oint_S \left(\nabla \times \vec{E} \right) \cdot d\vec{A} = 0. \tag{7.3}$$

The direction of the normal vector to the surface $d\vec{A}$, is determined according to the right-hand rule: if the fingers of the right hand curl in the direction of the path integral, the thumb points in the direction of $d\vec{A}$. The right-hand side can be rewritten using the zero vector $\vec{0}$,

$$\oint_S \left(\nabla \times \vec{E} \right) \cdot d\vec{A} = \oint_S \vec{0} \cdot d\vec{A}. \tag{7.4}$$

Rearranging the above equation, and using the distributive property of the dot product,

$$\oint_S \left(\nabla \times \vec{E} - \vec{0} \right) \cdot d\vec{A} = 0. \tag{7.5}$$

The above equation always holds true for any surface area S, and any vector field $\vec{E}$, if

$$\nabla \times \vec{E} = \vec{0}. \tag{7.6}$$

This is the differential form of Equation 7.2, which must be satisfied at every *point*, compared to the integral form in Equation 7.2, which must be satisfied over any loop.

7.2 The units of $\vec{D}$ and $\vec{E}$

The units of $[\vec{D}]$ and $[\vec{E}]$ are the same in ESU. This also holds true for $[\vec{B}]$ and $[\vec{H}]$ in EMU, which will be discussed in Chapter 11. For now, the focus is on $[\vec{D}]$ and $[\vec{E}]$.

Although there are many systems of units, ESU, EMU, SI, etc., they may differ in the mathematical formulation, but the underlying physics remain, and must remain the same. For example, it is possible to rewrite (not reformulate) Maxwell's equations, so that $[\vec{D}]$ and $[\vec{E}]$ (as well as $[\vec{B}]$ and $[\vec{H}]$) have the same units in the present SI formulation. This exercise has been done in Chapter 4, where the electric-displacement field $\vec{D}^*$, has been redefined without the constant ϵ_o, and therefore, having the same unit as the electric field $\vec{E}$. In some cases, electric displacement will be viewed as a different type of electric field: $\vec{D}_C$ in Chapter 15, and $\vec{D}'_F$ in Chapter 17.

8
Electric Displacement and Polarization Current in Ampere's Law

As formulated in Faraday's law, a time-varying magnetic-flux density $\vec{B}$ generates an electric field $\vec{E}$. By symmetry, a time-varying electric-flux density $\vec{D}$ generates a magnetic field $\vec{H}$, and captured in Ampere's law,

$$\nabla \times \vec{H} = \vec{J} + \frac{\partial \vec{D}}{\partial t}. \tag{8.1}$$

The above equation can be logically proven, by deriving the current-continuity equation from Ampere's law and Gauss's law [3]. From dimensional analysis, since $\frac{\partial \vec{D}}{\partial t}$ is added to the conduction current density $\vec{J}$, the unit of $[\frac{\partial \vec{D}}{\partial t}]$ must also be current density. $\frac{\partial \vec{D}}{\partial t}$ is known as the displacement current density, and can be viewed as a different type of "current density". This term can be shown to complete the current flow across the dielectric of a capacitor [3].

It was noted in Chapter 7, the net electric field that exists, including the effect of the dielectric material, is $\vec{E}$. In that case, why should Ampere's law be written using the electric displacement $\vec{D}$, instead of the electric field $\vec{E}$? This question will be answered in the remainder of this chapter.

Substituting Equation 6.1, which is equivalent to Equation 6.5, in Ampere's law,

$$\nabla \times \vec{H} = \vec{J} + \epsilon_o \frac{\partial \vec{E}}{\partial t} + \frac{\partial \vec{P}}{\partial t}. \tag{8.2}$$

The displacement current density can be written as the sum of two other current densities, in the above equation. As noted earlier, the two operands added to $\vec{J}$, must have the same unit as current density. The second operand is written using the electric field, which makes sense, since the net electric field that exists at any point, including the effect of the material at that point is $\vec{E}$.

Ampere's law written using $\vec{D}$, results in the third term $\frac{\partial \vec{P}}{\partial t}$. This term is known as the polarization current density. In Section 6.2, the polarization vector $\vec{P}$ is viewed as the average dipole moment per unit volume. The time-varying dipole moments of a dielectric material, caused by a time-varying electric field, are shown in Figure 8.1. In Figure 8.1(a), the electric field to the right, exerts a force on the positive charge in the atoms to the right, and the negative charge in the atoms to the left. This creates an electric dipole moment, a vector, whose direction is to the right, from the negative charge to the positive charge in the dipole.

If the time-varying electric field at the point (x_o, y_o, z_o) is a sinusoidal signal, for example,

$$\vec{E}(x_o, y_o, z_o, t) = a \sin(\omega t + \phi) \hat{x}, \tag{8.3}$$

the magnitude of the electric field, oscillates in a sinusoidal variation. If the electric field at its maximum amplitude is shown in Figure 8.1(a), the electric field at its minimum amplitude is shown in Figure 8.1(b), whose direction is now opposite, compared to its maximum amplitude. The electric dipole moment vector is also now in the opposite direction, and the positive charge is now skewed to the left, and the negative charge to the right. The term $\frac{\partial \vec{P}}{\partial t}$ signifies this motion of the dipole charges.

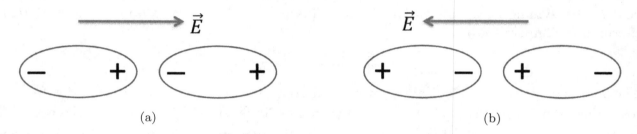

Figure 8.1: The polarization of atoms in a dielectric material, when the electric field is (a) to the right, and (b) to the left.

The motion of the dipole charges, by definition, is also a current, and generates a magnetic field, very much like the behavior of the typical current density $\vec{J}$. The current resulting from the time-varying dipole moment, are not due to charges that flow freely, rather due to the motion of the bound charges, and is called the bound current density. $\vec{J}$ in Ampere's law is sometimes referred to as the free current density, sometimes marked $\vec{J}_f$, since its the current density resulting from free charges that are free to flow in a conductor.

9

Introducing the Magnetic-Flux Density $\vec{B}$

Biot-Savart's law relates current i, and the magnetic field $\vec{H}$ due to the current. A detailed derivation of Biot-Savart's law is presented in Reference [3]. In SI units, Biot-Savart's law is

$$\vec{H} = \frac{1}{4\pi} \int_L \frac{i}{r^2} \left(\vec{dl} \times \hat{r} \right), \tag{9.1}$$

where i is the current in amperes, and the lengths are in meters. Biot-Savart's law is used to define the magnetic field $\vec{H}$.

The motivation for introducing a new electrical quantity, magnetic-flux density $\vec{B}$, can be understood from the ballistic-galvanometer experiment. A simplified version of this experiment is presented here. A detailed version is presented in Reference [3].

9.1 Ballistic-Galvanometer Experiment

The primary coil, marked P, in Figure 9.1 is wound around a toroid, and is connected to a battery B via switch W. The experiment is carried out on a toroid made of iron and wood, as examples of a magnetic and a non-magnetic material, respectively. A secondary coil S with many turns is wound over the primary coil in the highlighted region. S is connected to a ballistic galvanometer G, whose throw of the needle is a measure of the strength of the induced current.

A top view of the horizontal slice of the toroid is shown in Figure 9.2. The magnetic field within the toroid is shown by the dashed circle in Figure 9.2. The direction of the magnetic field, corresponds to a particular direction of the current in the coil, determined by the right-hand rule [3].

When the switch W is closed, the time-varying current in the primary coil sets up a time-varying magnetic field, which couples to the secondary coil, since S is wound over P in the shaded region in Figure 9.1. The time-varying magnetic flux in the secondary coil induces a transient current in the secondary coil, which can be detected from the throw of the needle in the ballistic galvanometer.

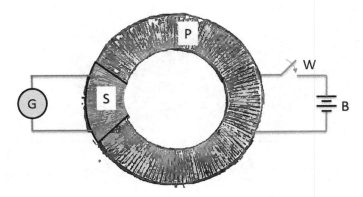

Figure 9.1: An experiment to show the effect of a magnetic material on an applied magnetic field

If the toroid is made up of a magnetic material such as iron, the transient current in the secondary coil when W is closed, captured by the throw of the galvanometer's needle, is different from the case when the toroid is made up of a non-magnetic material such as wood.

9.2 Magnetic-Flux Density $\vec{B}$

From the ballistic-galvanometer experiment, it can be concluded that the flux across the secondary coil must be different between a magnetic and a non-magnetic material. This shows that a magnetic material modifies the strength of the magnetic field.

The modified magnetic field at a point, by the material present at that point, is denoted as $\vec{B}$. Its early name is magnetic induction, since it is the field that generates an induced current caused by a time-varying magnetic flux. The field that would be present at a point, if the material at that point does not modify the applied magnetic field, is $\vec{H}$. The source of the applied magnetic field at a point can be from a current-carrying wire, as seen in the experiment described, and/or the magnetic field from a magnetized material near the point.

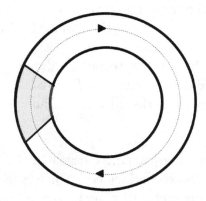

Figure 9.2: The magnetic field in the toroid in Figure 9.1.

$\vec{B}$ is written in EMU as a scaled factor of $\vec{H}$,

$$\vec{B} = \mu_r \vec{H}, \tag{9.2, EMU}$$

where μ_r is the relative permeability. μ_r has no units, and only scales $\vec{H}$. Therefore, $[\vec{B}]$ and $[\vec{H}]$ have the same units in EMU. This is similar to the relation between $\vec{D}$ and $\vec{E}$ in ESU, discussed earlier in Chapter 4.

In SI units, the relation between $\vec{B}$ and $\vec{H}$ is

$$\vec{B} = \mu_r \mu_o \vec{H}, \tag{9.3}$$

where μ_o is a constant with units, the permeability of free space, noted in Table 3.2. Therefore, the units of $[\vec{B}]$ and $[\vec{H}]$ are different. However, as shown in Chapter 4, it is possible to rewrite Maxwell's equations in SI units, such that $[\vec{B}]$ and $[\vec{H}]$ have the same units. One can view, therefore, $\vec{B}$ as a different type of magnetic field $\vec{H}$, as the magnetic field that includes the effect of the magnetic material.

10

Magnetization Vector $\vec{I}$

Similar to Equation 5.13 written as a sum in Equation 6.1, Equation 9.3 can be written as a sum of two vectors, rather than scaling $\vec{H}$ by a factor of $\mu_r \mu_o$,

$$\vec{B} = \mu_o \vec{H} + \mu_o \vec{I}, \tag{10.1}$$

where $\vec{I}$ is called the magnetization vector [3]. In the above equation, $\vec{H}$ is the magnetic field that would exist at a point, if the material at that point does not modify the field. The contribution of the material to the field at that point is $\vec{I}$, and the total field at the point $\vec{B}$, is the sum of the two, as written in the above equation, and scaled by the permeability of free space μ_o. As discussed in Chapter 4, it is possible to define $\vec{B}$ in SI, without the constant μ_o.

Note the asymmetry in how the constant μ_o is written in Equation 10.1, compared to the factor ϵ_o in Equation 6.1. ϵ_o is included as part of the definition of $\vec{P}$, but μ_o is not included as part of the definition of $\vec{I}$. This will be explained at the end of this chapter.

10.1 Magnetic Susceptibility

If the magnetization vector $\vec{I}$ is written as

$$\vec{I} = \chi_m \vec{H}, \tag{10.2}$$

where χ_m is the magnetic susceptibility, Equation 10.1 becomes

$$\vec{B} = \mu_o (1 + \chi_m) \vec{H}. \tag{10.3}$$

Equating the above relation to Equation 9.2,

$$\mu_r = 1 + \chi_m. \tag{10.4}$$

The advantage of writing the relative permeability in terms of the susceptibility, is that for values of μ_r that are approximately equal to 1, but not exactly equal to 1, such as 1.0000001 or 0.9999999, it is convenient to express the relative permeability in terms of the susceptibility.

10.2 The Definition of the Magnetic Dipole Moment With Magnetic Charges [Optional]

Similar to how the polarization vector $\vec{P}$ can be viewed as the electric dipole moment per unit volume, the magnetization vector $\vec{I}$ can be viewed as the magnetic dipole moment per unit volume. This will be discussed in Section 10.4, after the magnetic dipole moment is defined.

If magnetic charges do exist, the magnetic dipole moment can be defined using magnetic charges. This will be discussed first. Alternately, the magnetic dipole moment can be defined using a current loop, discussed in the following section.

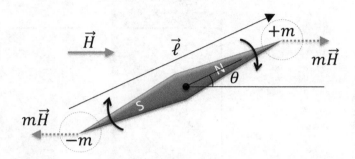

Figure 10.1: A magnetic needle modeled using magnetic charges $\pm m$.

The top view of a magnetic needle is shown in Figure 10.1, and is free to rotate on the plane of the page about its center pivot. The needle is modeled with magnetic charges $\pm m$ at its end points, $+m$ at the north pole N, and $-m$ at the south pole S. The direction of the terrestrial magnetic field is shown by the arrow marked $\vec{H}$. The magnetic charges experience a force

$$\vec{F} = \pm m \vec{H}, \tag{10.5}$$

which generates a torque on the needle, rotating the needle about its center pivot point, and aligning it to the magnetic field $\vec{H}$. If ℓ is the length of the needle, the magnitude of the torque on the needle is

$$\tau = 2 \times m \left(\frac{\ell}{2}\right) H \sin\theta. \tag{10.6}$$

The factor $2\times$ arises from the force acting on both the magnetic charges $\pm m$. The above equation is written as a cross product,

$$\vec{\tau} = m\vec{\ell} \times \vec{H}, \tag{10.7}$$

where $\vec{\ell}$ is the length vector in the direction from $-m$ to $+m$, as shown in the figure. This is the convention followed for the direction of $\vec{\ell}$, and the order of the operands in the cross product. Noting the similarity to Equation 6.16,

$$\vec{m} = m\vec{\ell}, \tag{10.8}$$

is known as the magnetic dipole moment.

10.3 The Magnetic Dipole Moment of a Current Loop

A rectangular loop of current i, shown in Figure 10.2, is used to define the magnetic dipole moment, instead of magnetic charges in Section 10.2. A rectangular loop is chosen for simplicity. Each of the sides are marked a–d. The current is assumed, for example, to flow in the direction marked by the arrows. The loop is free to pivot about its center, marked P, in any direction. The coordinate system is shown in the figure.

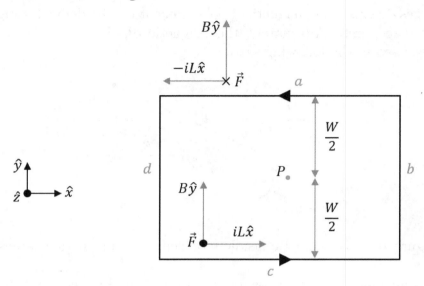

Figure 10.2: A rectangular current loop, instead of magnetic charges in Figure 10.1, is used to define the magnetic dipole moment.

The current loop is present in a constant magnetic-flux density $B\hat{y}$. The current loop is divided into small equal segments $\Delta\vec{l}$, and the direction of the vector is in the direction of the current flow in the segment. The force on each of the segments is

$$\Delta\vec{F} = i\Delta\vec{l} \times \vec{B}. \tag{10.9}$$

The force acting on each of the equal segments on a side is the same for all of the segments, since $i\Delta\vec{l}$ and $\vec{B}$ are the same vectors for all the segments on the same side. For the calculation of torque, by symmetry, the total force on the segments on a side, can be viewed as acting at the midpoint of the side. This is explained by an example.

Two segments, marked L and R, equidistant from the midpoint on Side c, is shown in Figure 10.3(a). The direction of the force acting on the segments in Equation 10.9 is out of the page, for the assumed directions of the current and the $\vec{B}$ field. The position vector $\vec{r}_L$, can be resolved into vectors $\vec{r}_x$ and $\vec{r}_y$, in the x and y directions. The torque $\vec{\tau}_L$ with respect to the pivot P, due to the force on Segment L is

$$\vec{\tau}_L = \vec{r}_L \times \Delta\vec{F} \tag{10.10}$$
$$= (\vec{r}_x + \vec{r}_y) \times \Delta\vec{F}. \tag{10.11}$$

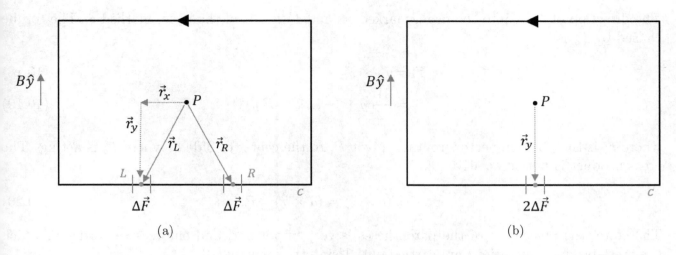

Figure 10.3: (a) Two tiny current elements along Side c, symmetric to the pivot P. (b) The force $\Delta \vec{F}$ acting on each of the segments L and R in Figure 10.3(a), is equivalent to the forces acting at the midpoint of Side c, for the calculation of the torque with respect to the pivot P.

Applying the distributive property of the cross product in the above equation,

$$\vec{\tau}_L = \vec{r}_x \times \Delta \vec{F} + \vec{r}_y \times \Delta \vec{F}. \tag{10.12}$$

By symmetry, the torque $\vec{\tau}_R$ due to the force on Segment R is

$$\vec{\tau}_R = -\vec{r}_x \times \Delta \vec{F} + \vec{r}_y \times \Delta \vec{F}. \tag{10.13}$$

The total torque due to the forces on the segments is the vector sum of the above two equations,

$$\vec{\tau}_L + \vec{\tau}_R = 2 \left(\vec{r}_y \times \Delta \vec{F} \right). \tag{10.14}$$

The above equation can be rewritten as

$$\vec{\tau}_L + \vec{\tau}_R = \vec{r}_y \times 2\Delta \vec{F}. \tag{10.15}$$

The above equation is the same as the torque resulting from the force $2\Delta \vec{F}$, acting at the midpoint of Side c, shown in Figure 10.3(b). This exercise shows that the total forces acting on the segments on the same side, can be viewed as acting at the midpoint of the side, for the calculation of the torque with respect to Point P.

Since $\Delta \vec{l}$ is parallel and anti-parallel to $B \hat{y}$ on the sides b and d, the force experienced by any of the segments on these two sides is 0. The total force on Side a, acting at the center of Side a is

$$\vec{F}_a = -iL\hat{x} \times B\hat{y}, \tag{10.16}$$

and the force on Side c is

$$\vec{F}_c = iL\hat{x} \times B\hat{y}. \tag{10.17}$$

The direction of $\vec{F}_a$ is into the page, marked ×, and $\vec{F}_c$ is out of the page, marked •. The torque caused by $\vec{F}_a$ is

$$\vec{\tau}_a = \vec{r}_a \times \vec{F}_a \tag{10.18}$$

$$= \frac{W}{2}\hat{y} \times (-iL\,\hat{x} \times B\,\hat{y}), \tag{10.19}$$

where $\vec{r}_a$ is the position vector from the pivot P, to the center of Side a, where $\vec{F}_a$ is acting. The cross product is not associative [9],

$$\vec{a} \times (\vec{b} \times \vec{c}) \neq (\vec{a} \times \vec{b}) \times \vec{c}. \tag{10.20}$$

Therefore, the placement of the parentheses is very important, and enforces the order in which the cross-product operations are carried out. Using the identity [10],

$$\vec{a} \times (\vec{b} \times \vec{c}) = (\vec{a} \times \vec{b}) \times \vec{c} + \vec{b} \times (\vec{a} \times \vec{c}), \tag{10.21}$$

Equation 10.19 can be written as

$$\vec{\tau}_a = \left[-\left(\frac{W}{2}\hat{y} \times iL\,\hat{x}\right) \times B\,\hat{y}\right] + \left[-iL\,\hat{x} \times \left(\frac{W}{2}\hat{y} \times B\,\hat{y}\right)\right]. \tag{10.22}$$

The second operand in the addition operation reduces to $\vec{0}$, since the cross product of parallel vectors is $\vec{0}$. Simplifying the expression,

$$\vec{\tau}_a = -\left(\frac{W}{2}\hat{y} \times iL\,\hat{x}\right) \times B\,\hat{y}. \tag{10.23}$$

Similarly, it is left as an exercise for the reader to verify

$$\vec{\tau}_c = -\left(\frac{W}{2}\hat{y} \times iL\,\hat{x}\right) \times B\,\hat{y}. \tag{10.24}$$

Using superposition, the total torque $\vec{\tau}$ is the vector sum of the above two equations,

$$\vec{\tau} = -\left(W\,\hat{y} \times iL\,\hat{x}\right) \times B\,\hat{y}. \tag{10.25}$$

Simplifying the above result,

$$\vec{\tau} = iA\,\hat{z} \times B\,\hat{y}, \tag{10.26}$$

where

$$A = WL \tag{10.27}$$

is the loop area. Equation 10.26 is similar to Equation 10.7, and

$$\vec{\eta} = iA\,\hat{z} \tag{10.28}$$

is defined as the magnetic dipole moment of the current loop. The magnetic dipole moment is now defined without the use of magnetic charges, unlike Equation 10.8. The magnetic dipole moment is a vector, whose direction is perpendicular to the loop area. This is clearly true in the case of the example presented. Although not rigorously proven mathematically, a qualitative proof is presented next.

It is shown in Reference [3], a current loop behaves like a bar magnet. The orientation of the magnet is perpendicular to the current loop, and the direction of the north pole can be determined by the right-hand rule: if the fingers of the right hand, curl around the loop in the direction of the current, the thumb points in the direction of the north pole.

The side view of a current loop is shown in Figure 10.4(a). The direction of the current in the loop is such that, the poles of the magnet that the current in the loop emulates, are as shown. As discussed in Section 10.2, the direction of the magnetic dipole moment of the bar magnet, marked $\vec{m}$, is from the south pole to the north. A similar example is shown in Figure 10.4(b), where the side view of the current loop is at an angle. The bar magnet is perpendicular to the current loop. For the direction of the current assumed, the poles of the magnet are as shown, determined by the right-hand rule stated earlier.

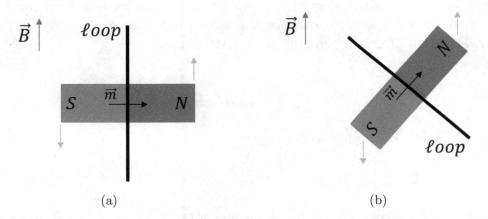

Figure 10.4: (a) The side view of a current loop, similar in behavior to the bar magnet. (b) The side view of the current loop at an angle.

Since the behavior of the current loop is similar to the bar magnet, the direction of the magnetic dipole moment of the current loop must also be perpendicular to the current loop. A similar right-hand rule for the direction of the magnetic dipole moment of the current loop, must therefore apply: if the fingers of the right hand, curl in the direction of the current in the loop, the thumb points in the direction of the magnetic dipole moment of the current loop. In the example presented in Figure 10.2, note that this rule is in agreement with the result in Equation 10.28.

In general, the magnetic dipole moment of a current loop in Equation 10.28 can be written

as
$$\vec{\eta} = iA\hat{n}, \tag{10.29}$$
where $\hat{n}$ is the unit normal vector to the planar loop area, and its direction is determined using the right-hand rule stated in the previous paragraph.

Since the current loop pivots about its center P, the equivalent magnet representing the loop, also pivots about its center. The forces on the poles of the magnet due to $\vec{B}$, are shown by the arrows marked near the magnet in Figure 10.4. The force on the magnet results in a torque, and causing the magnet to rotate and aligning to $\vec{B}$, similar to the magnetic compass in Section 10.2. Since the magnetic dipole moment vector of the current loop $\vec{\eta}$, is in the same direction as the magnet, the torque on the current loop causes the magnetic dipole moment vector of the current loop to align to $\vec{B}$.

The representation of the current-carrying loop by a magnet, is not specific to a rectangular loop, but also holds true for a circular loop, for example. Although not rigorously mathematically proven, one can expect Equation 10.29 to be valid for any shape of the current loop of area A.

There is a torque on the magnet representing the current loop, aligning the magnet or the dipole moment of the current loop to $\vec{B}$, at any angular offset of the current loop from $\vec{B}$. Similar to how Equation 10.7 is valid for any angle between the magnetic needle and $\vec{H}$, a hypothesis can be made that Equation 10.26, in general, can be written as

$$\vec{\tau} = \vec{\eta} \times \vec{B} \tag{10.30}$$
$$= iA\hat{n} \times \vec{B}, \tag{10.31}$$

holds true for any angle between $\hat{n}$ and $\vec{B}$.

10.4 Magnetization Vector $\vec{I}$

In ESU, the relation between the electric displacement $\vec{D}$, the electric field $\vec{E}$, and the polarization vector $\vec{P}$ is
$$\vec{D} = \vec{E} + 4\pi\vec{P}. \tag{10.32, ESU}$$
This equation can also be written as
$$\vec{D} = \epsilon_r \vec{E}, \tag{10.33, ESU}$$
where ϵ_r is the relative permittivity. In EMU, a similar relation applies between the magnetic-flux density $\vec{B}$, the magnetic field $\vec{H}$, and the magnetization vector $\vec{I}$,
$$\vec{B} = \vec{H} + 4\pi\vec{I}. \tag{10.34, EMU}$$
The above equation can also be written as
$$\vec{B} = \mu_r \vec{H}, \tag{10.35, EMU}$$

where μ_r is the relative permeability. The above equations are derived and explained in great detail in Reference [3].

From these equations, it can be seen that $(\vec{B}, \vec{D})$, $(\vec{H}, \vec{E})$, and $(\vec{I}, \vec{P})$ are symmetrical from the perspective of a mathematical formulation. It was shown earlier that the polarization vector $\vec{P}$ can be defined as the average electric dipole moment per unit volume. Analogous to the definition of $\vec{P}$, $\vec{I}$ can be viewed as the average magnetic dipole moment per unit volume. In SI units, Equation 10.34 is written as Equation 10.1.

The average magnetic dipole moment per unit volume, similar to Equation 6.26, can be rigorously defined as

$$\vec{I} = \frac{\frac{1}{N} \sum_{i=1}^{N} \vec{\eta}_i}{\Delta V}, \tag{10.36}$$

where $\vec{\eta}_i$ is the ith dipole moment of a tiny current loop of area A, with the unit normal vector $\hat{n}_i$, determined according to the right-hand rule in Section 10.3,

$$\vec{\eta}_i = iA\,\hat{n}_i. \tag{10.37}$$

N is the number of current loops in volume ΔV. Of course, the above equation doesn't mean that one can count the number of current loops, and compute the average dipole moment value in a volume to calculate $\vec{I}$. Equation 10.36 is a different interpretation of the magnetization vector $\vec{I}$.

Applying dimensional analysis in the above equation, note that the SI unit of the magnetic dipole moment is

$$[\vec{\eta}] = A\,m^2. \tag{10.38}$$

Using this result in Equation 10.36, the unit of the magnetization vector is

$$[\vec{I}] = \frac{A}{m}, \tag{10.39}$$

which is the same as the SI unit of the magnetic field $\vec{H}$, from Table 3.1. This is consistent with Equation 10.1, which requires the units of $\vec{H}$ and $\vec{I}$ to be the same, since both are scaled by the same constant μ_o, and added together. This is the reason for not defining the magnetization vector, including the constant μ_o, which is a break in the symmetry between Equation 10.1 and Equation 6.1. If μ_o, a constant with units in Table 3.2, is included as part of the definition of $\vec{I}$, then $\vec{I}$ can no longer be interpreted as the magnetic dipole moment per unit volume, since the units will not be correct.

11

Key Points on Magnetic Field $\vec{H}$ and Magnetic-Flux Density $\vec{B}$

Unlike Equation 3.6, it is possible to rewrite Maxwell's equations, and define the relation between $\vec{B}$ and $\vec{H}$, without the constant μ_o, similar to EMU. Repeating Equation 4.8,

$$\vec{B}^* = \mu_r \vec{H}. \tag{11.1}$$

$\vec{B}^*$ is used instead of $\vec{B}$, to indicate that the magnetic-flux density is defined without the constant μ_o. Since μ_r has no units, $[\vec{B}^*]$ and $[\vec{H}]$ have the same units. In this regard, ignoring the constant μ_o, $\vec{B}$ can be viewed as a different type of magnetic field.

$\vec{B}$ (or $\vec{B}^*$) is the net magnetic field at any point, which includes the effect of the magnetic material at that point, on an applied magnetic field $\vec{H}$. This is similar to how the electric field $\vec{E}$ is the net electric field, including the effect of the dielectric material, discussed earlier in Chapter 7.

12

On the Formulation of Ampere's Law

Ampere's law can be viewed as a cause-effect relation, where the conduction current density term $\vec{J}$, and/or the displacement current density $\frac{\partial \vec{D}}{\partial t}$, cause the effect of generating the magnetic field $\vec{H}$,

$$\nabla \times \vec{H} = \vec{J} + \frac{\partial \vec{D}}{\partial t}. \tag{12.1}$$

Ampere's law is formulated using the magnetic field $\vec{H}$, and not the magnetic-flux density $\vec{B}^*$. As discussed in Chapter 4, $\vec{B}^*$ has the same unit as $\vec{H}$. For the discussion in this chapter, $\vec{B}^*$ can be swapped for $\vec{H}$, without affecting the units.

The focus of this chapter is to show that the formulation of Ampere's law using magnetic-flux density $\vec{B}^*$, swapping $\vec{H}$ for $\vec{B}^*$, the left-hand side of the above equation is modified as

$$\nabla \times \vec{B}^*, \tag{12.2}$$

results in a non-physical field solution. A similar analysis will be repeated in Chapter 17, to make a hypothesis, and show that Faraday's law can be written in a more general way, using a new electric-displacement field, instead of the electric field.

A simple thought experiment is presented next. Two cases are shown in Figure 12.1: In Case 1, for simplicity, a uniform isotropic linear material of relative permeability μ_{r1} is present, while a different material of relative permeability μ_{r2} in Case 2. The same conduction current density and displacement current density are present in both the cases.

Each of the Maxwell's equations must be meaningful, when analyzed independently, although Faraday's law and Ampere's law are coupled to each other. A time-varying current source in Ampere's law, generates a time-varying magnetic field. This effect is coupled to Faraday's law. A time-varying magnetic field, generates a time-varying electric field, which results in a magnetic field in Ampere's law, and so on. However, when Ampere's law is analyzed independently, the formulation of this equation must make sense.

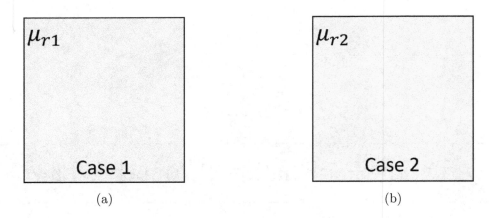

Figure 12.1: A uniform magnetic material of relative permeability (a) μ_{r1}, and (b) μ_{r2}.

Alternately, in the thought experiment, it can be assumed that the conduction and the displacement current densities are static, and not time varying, resulting in a static magnetic field. If the conduction current density $\vec{J}$ is a constant, and if the electric-flux density $\vec{D}$ is a linear function of time, then the displacement current density is a constant, and this results in a static magnetic field in Equation 12.1. Now, Ampere's law is no longer coupled to Faraday's law, since

$$\frac{\partial \vec{B}}{\partial t} = 0, \tag{12.3}$$

and Ampere's law can be analyzed independently.

If the left-hand side of Equation 12.1 is modified as Equation 12.2, the field $\vec{B}^*$ generated in both the cases are equal, since the sources in the right-hand side of the equation are identical in both the cases. The magnetic field $\vec{H}$, however, are different. Applying Equation 4.8, the magnetic field $\vec{H}_1$ in Case 1,

$$\vec{H}_1 = \frac{\vec{B}^*}{\mu_{r1}}, \tag{12.4}$$

is not equal to the magnetic field $\vec{H}_2$ in Case 2,

$$\vec{H}_2 = \frac{\vec{B}^*}{\mu_{r2}} \neq \vec{H}_1, \tag{12.5}$$

since the relative permeability values are different in both the cases.

This is a non-physical result! As discussed in Chapter 11, by definition, $\vec{B}^*$ is the field that exists, including the effect of the magnetic material. The following two steps occur in the generation of $\vec{B}^*$, in this order:

1. The current sources generate the magnetic field $\vec{H}$.
2. The effect of a magnetic material on $\vec{H}$ modifies this field, resulting in $\vec{B}^*$.

In the non-physical result, two different magnetic fields, $\vec{H}_1$ and $\vec{H}_2$, are generated by identical current density sources, and the effect of the material is to modify this field, resulting in the *same* field $\vec{B}^*$. Since the current density sources are identical in both the cases, in the formulation using Equation 12.2, where

$$\vec{H}_1 \neq \vec{H}_2, \qquad (12.6)$$

makes no physical sense.

This simple exercise, verifies that Ampere's law in Equation 12.1, is indeed the correct formulation. In this case, the same magnetic field $\vec{H}$ is generated in both the cases, since the source terms on the right-hand side are identical in both the cases. The magnetic-flux density in Case 1 $\vec{B}_1^*$,

$$\vec{B}_1^* = \mu_{r1}\vec{H}, \qquad (12.7)$$

however, is not equal to the magnetic-flux density $\vec{B}_2^*$ in Case 2,

$$\vec{B}_2^* = \mu_{r2}\vec{H} \neq \vec{B}_1^*, \qquad (12.8)$$

and makes physical sense, since the magnetic materials modifying $\vec{H}$ to $\vec{B}$, are different in both the cases.

13

Rewriting Maxwell's Equations With Electric Field Separated by its Sources

In this chapter, Maxwell's equations will be rewritten, not reformulated, in a different form. The electric field and the electric-displacement field will be written as a superposition of fields, based on the source of the field. This will be required for the reformulation of Faraday's law, explained in Chapter 18.

13.1 Sources of Electric Displacement $\vec{D}$ and Electric Field $\vec{E}$

There are two sources of electric field $\vec{E}$, and from the definition of $\vec{D}$ in Equation 5.13, these two sources are applicable to the sources of $\vec{D}$ as well:

1. Electric charges q: Its relation to electric field is captured in Gauss's law. The electric displacement and electric fields associated with electric charges, will be referred to as $\vec{D}_C$ and $\vec{E}_C$.

2. Time-varying magnetic-flux density $\vec{B}$: The generation of the electric field due to a time-varying magnetic-flux density $\vec{B}$ is captured in Faraday's law. The electric displacement and electric fields associated with the time-varying magnetic-flux density, will be referred to as $\vec{D}_F$ and $\vec{E}_F$.

There is a third source of electric field: magnetic currents [3]. However, since magnetic currents exist only in theory, this source will be ignored.

13.2 Rewriting Maxwell's Equations

The magnetic vector potential $\vec{A}$ is defined as

$$\vec{B} = \nabla \times \vec{A}, \tag{13.1}$$

meeting the divergence-free condition of $\vec{B}$ in Equation 3.3, since the divergence of curl of a vector field is 0. Substituting the above equation in the present formulation of Faraday's law,

$$\nabla \times \vec{E} = -\frac{\partial}{\partial t}\left(\nabla \times \vec{A}\right). \tag{13.2}$$

Rearranging the above equation,

$$\nabla \times \left(\vec{E} + \frac{\partial \vec{A}}{\partial t}\right) = 0. \tag{13.3}$$

From calculus, if the curl of a vector field is 0, the vector field can be written as the gradient of a scalar field,

$$\vec{E} + \frac{\partial \vec{A}}{\partial t} = -\nabla \Phi. \tag{13.4}$$

Rearranging the above equation,

$$\vec{E} = -\nabla \Phi - \frac{\partial \vec{A}}{\partial t} \tag{13.5}$$

$$= \vec{E}_C + \vec{E}_F. \tag{13.6}$$

Intuitively, the above equation is the superposition of the total electric field at any point, as a sum of the electric field $\vec{E}_C$ due to electric charges, and the electric field $\vec{E}_F$ generated from a time-varying magnetic flux, captured in Faraday's law,

$$\vec{E}_C = -\nabla \Phi, \tag{13.7}$$

and

$$\vec{E}_F = -\frac{\partial \vec{A}}{\partial t}. \tag{13.8}$$

The above equation is the same as Faraday's law, disguised in a different form. Applying the $\nabla \times$ operator on both sides of the above equation, and substituting Equation 13.1,

$$\nabla \times \vec{E}_F = -\frac{\partial \vec{B}}{\partial t}, \tag{13.9}$$

which is the same as Faraday's law. Intuitively, it can be noted, since $\vec{E}_F$ is the electric field generated by Faraday's law, the other source of electric field $\vec{E}_C$, must be the electric field generated by electric charges. This can be proven mathematically using the Coulomb gauge,

$$\nabla \cdot \vec{A} = 0, \tag{13.10}$$

to show that $\vec{E}_C$ is the instantaneous electric field due to electric charges [11]–[12]. The electric field $\vec{E}_C$ will be referred to as the Coulomb electric field, since its the field due to electric charges. The electric field $\vec{E}_F$ will be referred to as the Faraday electric field, since its the electric field generated in Faraday's law.

Since $\vec{E}_C$ is the gradient of a scalar potential in Equation 13.7, from calculus,
$$\nabla \times \vec{E}_C = 0, \tag{13.11}$$
at any time instant. Multiplying Equation 13.6 by $\epsilon_r \epsilon_o$, and substituting Equation 3.5,
$$\vec{D} = \vec{D}_C + \vec{D}_F, \tag{13.12}$$
where
$$\vec{D}_C = \epsilon_r \epsilon_o \vec{E}_C \tag{13.13}$$
$$\vec{D}_F = \epsilon_r \epsilon_o \vec{E}_F. \tag{13.14}$$

Similar to Equation 13.6, $\vec{D}$ at any point can be written as Equation 13.12. Using the above equation, Ampere's law is written as
$$\nabla \times \vec{H} = \vec{J} + \frac{\partial}{\partial t}\left(\vec{D}_C + \vec{D}_F\right). \tag{13.15}$$
Applying the $\nabla \cdot$ operation on both the sides of the above equation,
$$\nabla \cdot \left(\nabla \times \vec{H}\right) = \nabla \cdot \vec{J} + \frac{\partial}{\partial t}\nabla \cdot \left(\vec{D}_C + \vec{D}_F\right). \tag{13.16}$$
From calculus, the left-hand side reduces to 0. If
$$\nabla \cdot \left(\vec{D}_C + \vec{D}_F\right) = \rho, \tag{13.17}$$
Equation 13.16 reduces to the current-continuity equation,
$$\nabla \cdot \vec{J} = -\frac{\partial \rho}{\partial t}. \tag{13.18}$$
This shows that
$$\nabla \cdot \left(\vec{D}_C + \vec{D}_F\right) = \rho, \tag{13.19}$$
or Gauss's law, must be satisfied for time-varying fields, at any time instant. The existing set of Maxwell's equations can be written as the following set of equations:

$$\nabla \cdot \left(\vec{D}_C + \vec{D}_F\right) = \rho \tag{13.20}$$
$$\nabla \cdot \vec{B} = 0 \tag{13.21}$$
$$\nabla \times \vec{E}_F = -\frac{\partial \vec{B}}{\partial t} \tag{13.22}$$
$$\nabla \times \vec{E}_C = \vec{0} \tag{13.23}$$
$$\nabla \times \vec{H} = \vec{J} + \frac{\partial}{\partial t}\left(\vec{D}_C + \vec{D}_F\right), \tag{13.24}$$

and the above equations are also valid for time-varying fields [3]. In addition to the above equations,

$$\vec{D}_F = \epsilon_r \epsilon_o \vec{E}_F \tag{13.25}$$
$$\vec{D}_C = \epsilon_r \epsilon_o \vec{E}_C \tag{13.26}$$
$$\vec{B} = \mu_r \mu_o \vec{H}, \tag{13.27}$$

capture the effect of a material on the fields. From Equation 6.1 and Equation 10.1, the above equations can also be written as

$$\vec{D}_F = \epsilon_o \vec{E}_F + \vec{P}_F \tag{13.28}$$
$$\vec{D}_C = \epsilon_o \vec{E}_C + \vec{P}_C \tag{13.29}$$
$$\vec{B} = \mu_o \vec{H} + \mu_o \vec{I}, \tag{13.30}$$

where $\vec{P}_C$ and $\vec{P}_F$ are the polarization vectors, corresponding to the electric displacement vectors $\vec{D}_C$ and $\vec{D}_F$. The values and units of the constants ϵ_o and μ_o are documented in Table 3.2. The electric displacement and electric field at any point, repeating Equation 13.12 and Equation 13.6, are

$$\vec{D} = \vec{D}_C + \vec{D}_F \tag{13.31}$$
$$\vec{E} = \vec{E}_C + \vec{E}_F. \tag{13.32}$$

14

Redefining $\vec{D}_C/\vec{D}_F$, $\vec{B}$, and $\vec{P}_C/\vec{P}_F$, Without the Constants ϵ_o and μ_o

Following Chapter 4, the electric displacement fields $\vec{D}_C/\vec{D}_F$, the magnetic-flux density $\vec{B}$, and the polarization vectors $\vec{P}_C/\vec{P}_F$, can be written without the constant ϵ_o or μ_o. This is useful for the discussion in future chapters.

The convention followed is that the variables defined without the constant ϵ_o or μ_o, are written with an asterisk superscript. Replacing the variables to the left of the arrow $\rightarrow$ by the ones on the right, the electromagnetic equations can be written with electric displacement vectors, polarization vectors, and magnetic-flux density, without the constants ϵ_o or μ_o,

$$\vec{D}_C \rightarrow \epsilon_o \vec{D}_C^*, \text{ where } \vec{D}_C^* = \epsilon_r \vec{E}_C \qquad (14.1)$$

$$\vec{D}_F \rightarrow \epsilon_o \vec{D}_F^*, \text{ where } \vec{D}_F^* = \epsilon_r \vec{E}_F \qquad (14.2)$$

$$\vec{B} \rightarrow \mu_o \vec{B}^*, \text{ where } \vec{B}^* = \mu_r \vec{H} \qquad (14.3)$$

$$\vec{P}_C \rightarrow \epsilon_o \vec{P}_C^*, \text{ where } \vec{P}_C^* = \chi_e \vec{E}_C \qquad (14.4)$$

$$\vec{P}_F \rightarrow \epsilon_o \vec{P}_F^*, \text{ where } \vec{P}_F^* = \chi_e \vec{E}_F. \qquad (14.5)$$

Replacing $\vec{D}_C$, $\vec{D}_F$, and $\vec{B}$ in Equation 13.28 – Equation 13.30 by the above expressions,

$$\vec{D}_C^* = \vec{E}_C + \vec{P}_C^* \qquad (14.6)$$

$$\vec{D}_F^* = \vec{E}_F + \vec{P}_F^* \qquad (14.7)$$

$$\vec{B}^* = \vec{H} + \vec{I}. \qquad (14.8)$$

As discussed in Chapter 10, note that the magnetization vector $\vec{I}$ is already defined without the constant μ_o, unlike the polarization vector $\vec{P}$, which is defined including the constant ϵ_o.

Applying dimensional analysis on the above equations, the electric-displacement fields $\vec{D}_C^*/\vec{D}_F^*$, the electric fields $\vec{E}_C/\vec{E}_F$, and the polarization vectors $\vec{P}_C^*/\vec{P}_F^*$ have the same units. Likewise, the magnetic-flux density $\vec{B}^*$, the magnetic field $\vec{H}$, and the magnetization vector $\vec{I}$ have the same units.

Substituting Equation 4.5 and Equation 14.1 – Equation 14.2 in Equation 13.31,

$$\vec{D}^* = \vec{D}_C^* + \vec{D}_F^*. \tag{14.9}$$

Applying dimensional analysis on the above equation, $\vec{D}_C^*$ and $\vec{D}_F^*$ have the same units as $\vec{D}^*$, documented in Table 4.1.

Note that the redefined polarization vectors $\vec{P}_C^*$ and $\vec{P}_F^*$, can no longer be viewed as the average dipole moment per unit volume, since the units are not consistent between the two. $\vec{P}_C^*$ and $\vec{P}_F^*$ will be viewed strictly from a mathematical perspective, as vectors to rewrite

$$\vec{D}_C^* = \epsilon_r \vec{E}_C \tag{14.10}$$
$$\vec{D}_F^* = \epsilon_r \vec{E}_F, \tag{14.11}$$

as a vector sum in Equation 14.6 – Equation 14.7, instead of the product form written above.

From the above equations, Ampere's law in Equation 13.24, for example, can be written as

$$\nabla \times \vec{H} = \vec{J} + \epsilon_o \frac{\partial}{\partial t} \left(\vec{D}_C^* + \vec{D}_F^* \right). \tag{14.12}$$

Substituting Equation 14.6 and Equation 14.7 in the above equation,

$$\nabla \times \vec{H} = \vec{J} + \epsilon_o \frac{\partial}{\partial t} \left(\vec{E}_C + \vec{P}_C^* \right) + \epsilon_o \frac{\partial}{\partial t} \left(\vec{E}_F + \vec{P}_F^* \right) \tag{14.13}$$
$$= \vec{J} + \vec{J}_C + \vec{J}_F, \tag{14.14}$$

where,

$$\vec{J}_C = \epsilon_o \frac{\partial \vec{D}_C^*}{\partial t}$$
$$= \epsilon_o \frac{\partial}{\partial t} \left(\vec{E}_C + \vec{P}_C^* \right) \tag{14.15}$$

is the displacement current density (see Chapter 8) associated with the electric-displacement field $\vec{D}_C^*$, due to electric charges, and

$$\vec{J}_F = \epsilon_o \frac{\partial \vec{D}_F^*}{\partial t}$$
$$= \epsilon_o \frac{\partial}{\partial t} \left(\vec{E}_F + \vec{P}_F^* \right), \tag{14.16}$$

is the displacement current density associated with the electric-displacement field $\vec{D}_F^*$, due to a time-varying magnetic-flux density in Faraday's law.

Gauss's law in Equation 13.20, for example, can be written as

$$\nabla \cdot \left(\epsilon_o \vec{D}_C^* + \epsilon_o \vec{D}_F^*\right) = \rho. \tag{14.17}$$

Substituting Equation 14.9 in the above equation,

$$\nabla \cdot \left(\epsilon_o \vec{D}^*\right) = \rho. \tag{14.18}$$

15
A New Perspective on the Meaning of Electric Displacement $\vec{D}_C$

Since the electric charges in Faraday's experiment with spherical capacitors are stationary, this is an experiment in *electrostatics*. The source of the electric field in the experiment are electric charges.

As discussed in Chapter 13, a subscript C will be added to the electric field and the electric-displacement vectors, as a reminder that the source of the fields are electric charges. Equation 5.8 will be rewritten as

$$\vec{E}_C = \frac{\vec{E}_{C,air}}{\epsilon_r}, \tag{15.1}$$

where $\vec{E}_C$ is the electric field at any point in a material medium. Similarly, Equation 5.13 that was derived for a material medium, will be written as

$$\vec{D}_C = \epsilon_r \epsilon_o \vec{E}_C, \tag{15.2}$$

since the derivation of the above equations, have been the result of the analysis of the electric field due to electric charges. A new meaning of electric displacement emerges, when Equation 15.1 is substituted in Equation 15.2, resulting in

$$\vec{D}_C = \epsilon_o \vec{E}_{C,air}. \tag{15.3}$$

The scaling factor of ϵ_o is the result of how the equations are organized in the SI system of units. Repeating Equation 14.1,

$$\vec{D}_C^* = \epsilon_r \vec{E}_C, \tag{15.4}$$

where $\vec{D}_C^*$ does not include ϵ_o as part of its definition, unlike Equation 15.2. This is useful for the discussion that follows, without concerning about the constant ϵ_o. As noted in Chapter 14, ϵ_o is now present in other equations, and different set of units than SI. The above equation is the same as how its written in ESU. Since ϵ_r is a dimensionless value, $\vec{D}$ and $\vec{E}$ have the same units, similar to ESU.

In this case, substituting Equation 15.1 in Equation 15.4, Equation 15.3 is now written as

$$\vec{D}_C^* = \vec{E}_{C,air}. \tag{15.5}$$

The above equation states that the electric displacement at any point, is the same as the electric field in air, or in other words, the electric displacement at any point is the same as the electric field that is not "modified", "altered", or reduced in strength by the dielectric material at that point. This perspective also holds true in the case of Equation 15.3: the electric displacement is the same as the electric field unmodified in a dielectric material, and scaled by the constant ϵ_o. Note that the electric field $\vec{E}_C$ at any point, by definition, as explained in Chapter 7, is the net electric field, including the effect of the dielectric material at that point.

The new perspective of $\vec{D}_C$ in the above paragraph, is specific to the electric displacement due to electric charges. This has been emphasized using the subscript C. In the case of the electric displacement due to a time-varying magnetic-flux density, a new electric-displacement field $\vec{D}_F'$ will be defined in Chapter 17, and carries a similar meaning.

The new perspective of $\vec{D}_C$ presented here, has been derived from the analysis of electric field in Faraday's experiment with spherical capacitors, in the following conditions:

1. The electric field is generated by electric charges.
2. The electric field is static, and not a time-varying field.
3. The electric field is present in a uniform dielectric material.

It will be assumed that the new perspective of $\vec{D}_C$ also holds true in general, in the case of time-varying fields, and a non-uniform material region. The purpose of this chapter is to associate a simple, and an easy to understand physical meaning with Equation 15.2.

16

On the Formulation of Faraday's Law

Faraday's law, repeating Equation 13.22,

$$\nabla \times \vec{E}_F = -\frac{\partial \vec{B}}{\partial t}, \qquad (16.1)$$

is analyzed in detail in this chapter. The above differential equation is valid at any point. One can most certainly view the above equation as a cause-effect relation: the cause is a time-varying magnetic-flux density $\vec{B}$, resulting in the effect of the electric field $\vec{E}_F$ generated. The differential form of Faraday's law, written in the above equation, is valid at any point in space.

Faraday's experiments with spherical capacitors, as noted in Chapter 5 and Chapter 7, clearly show that a dielectric material at a point has an effect on the electric field at that point, and modifying the field. There are only 2 possibilities in regard to the effect of a dielectric material on the electric field generated by a time-varying $\vec{B}$:

(I) The dielectric material has a non-NULL effect, modifying the electric field generated by a time-varying magnetic-flux density $\vec{B}$.

(II) The dielectric material has a NULL effect, where the dielectric material has absolutely NO effect in modifying the electric field.

Each of the above two cases is discussed in more detail next.

16.1 Case (I): non-NULL effect

There are two possibilities in how the non-NULL effect of a dielectric material, on the electric field generated by a time-varying magnetic-flux density $\vec{B}$, can be captured in Faraday's law. These will be referred to as Case A and Case B discussed in the following two subsections.

16.1.1 Case A

Faraday's law can be formulated such that $\vec{E}_F$ in Equation 16.1, *includes* the effect of the dielectric material, and is the modified electric field, generated by a time-varying magnetic-flux density $\vec{B}$.

This case is consistent with the definition of electric field $\vec{E}$. As noted in Chapter 7, by definition, $\vec{E}$ represents the electric field at any point, that includes the effect of the dielectric material at that point.

This is a non-physical formulation of Faraday's law! This will be shown using a thought experiment in Chapter 17. Even without a thought experiment, it can be noted that in a cause-effect relation, the time-varying magnetic-flux density at any point, does not know what material is present at that point, to generate the correct electric field that includes the effect of the dielectric material.

16.1.2 Case B

Since Case A is non-physical, Faraday's law must be formulated such that the electric field generated by a time-varying magnetic-flux density $\vec{B}$, does not include the effect of the dielectric material. Faraday's law will be reformulated in Chapter 17, using a new electric-displacement field, instead of the electric field.

16.2 Case (II): NULL effect

In this case, the dielectric material has NO effect on the electric field generated from a time-varying magnetic-flux density $\vec{B}$. This case will be referred to as Case C. Although Faraday's experiments with spherical capacitors show that a dielectric material modifies the electric field due to electric charges, its possible that this case can happen! The electric field generated by a time-varying $\vec{B}$ is of a different mechanism, compared to the electric field due to electric charges. This case, however, appears to be too ideal.

In Chapter 17, Case C will be shown to be a special case of Case B, and the reformulated Faraday's law according to Case B, reduces to the existing formulation of Faraday's law in Equation 16.1. If all the different dielectric materials that exist in the universe, has absolutely no effect on the electric field generated from a time-varying magnetic-flux density $\vec{B}$, Faraday's law requires no reformulation. Since Case C appears to be too ideal, a hypothesis will be made, and Faraday's law will be reformulated according to Case B.

17

The Reformulation of Faraday's Law Using Electric Displacement Instead of Electric Field

Case A and Case B in Chapter 16, are two ways by which Faraday's law can be formulated, to account for the non-NULL effect of a dielectric material, on the electric field generated by a time-varying magnetic-flux density $\vec{B}$. In Case A, the electric field $\vec{E}_F$ at any point in Faraday's law,

$$\nabla \times \vec{E}_F = -\frac{\partial \vec{B}}{\partial t}, \qquad (17.1)$$

includes the effect of the dielectric material on the electric field at that point.

Faraday's law can be viewed as a cause-effect relation: the cause of a time-varying $\vec{B}$ in the right-hand side of the above equation, results in the effect of the electric field generated. The time-varying $\vec{B}$ does not know what dielectric material is present, to generate the correct electric field, which includes the effect of the dielectric material. Even without a thought experiment, Case A makes no physical sense.

Using a similar thought experiment in Chapter 12, it will be shown that Case A is a non-physical formulation. Faraday's law will be reformulated according to Case B, discussed in Section 17.2, following which, Case C of Chapter 16, will be shown to be a special case of Case B.

17.1 Introducing a New Permittivity κ of the Faraday Electric Field

Faraday's experiments with spherical capacitors in Chapter 5, show that a dielectric material at any point has an effect on the electric field at that point, modifying the field. In Case A, since $\vec{E}_F$ in Faraday's law includes the effect of the dielectric material at any point, a new permittivity κ is introduced, so that

$$\vec{D}'_F = \kappa \vec{E}_F, \qquad (17.2)$$

is the electric field at a point, without the effect of the dielectric material at that point. The above equation bears resemblance to Equation 14.2,

$$\vec{D}^*_F = \epsilon_r \vec{E}_F. \qquad (17.3)$$

In Chapter 21, it will be shown that
$$\kappa \neq \epsilon_r. \tag{17.4}$$
In the experiment results with distilled water in Chapter 22, although its $\epsilon_r \approx 78$ value is high,
$$\kappa \approx 1, \tag{17.5}$$
but not precisely equal to 1.0. A preview of Chapter 21, the reason for introducing a new permittivity κ, is a consequence of the difference in the behavior of the Faraday electric field and the Coulomb electric field in a dielectric medium. This shows that it takes two parameters, κ and ϵ_r, to characterize a dielectric material, and not just ϵ_r.

ϵ_r captures the effect of a dielectric material on the electric field $\vec{D}_C^*$ due to electric charges, formulated in Equation 5.8. ϵ_r will be referred to as the (relative) permittivity of the Coulomb electric field, to differentiate between the two kinds of permittivity. κ captures the effect of a dielectric material on the electric field $\vec{D}_F'$, due to a time-varying magnetic-flux density $\vec{B}$. For this reason, κ will be referred to as the permittivity of the Faraday electric field.

Similar to the permittivity tensor [3], κ is also represented as a tensor in complex materials, such as an anisotropic material.

17.2 A Thought Experiment Similar to Chapter 12

Two cases are analyzed: the material in Case 1 in Figure 17.1(a) is a uniform isotropic dielectric material of the permittivity of the Faraday electric field κ_1, while the region in Case 2 in Figure 17.1(b) is filled with a uniform isotropic material of the permittivity of the Faraday electric field κ_2. A uniform isotropic dielectric material is assumed for simplicity, and is sufficient for presenting the motivation to modify Faraday's law. The reformulated Maxwell's equations, however, is also valid for any type of material, such as an anisotropic or a non-linear material, no different from the existing formulation.

Each of the Maxwell's equations must be meaningful when analyzed on its own, although Faraday's law and Ampere's law are coupled together, as discussed in Chapter 12. Alternately, in the thought experiment, one could think of $\vec{B}$ as linearly varying, resulting in a static or a non time-varying electric field. Now, Faraday's law is no longer coupled to Ampere's law, and can be analyzed independently.

The time-varying magnetic-flux density $\vec{B}$ is identical in both the cases. Case 1 and Case 2 are identical, except for the permittivity of the material that fills the space. Since $\vec{B}$ is identical in both the cases, by Faraday's law, the electric field $\vec{E}_F$ generated in each of the cases must be equal as well.

In Case A, the electric field in Faraday's law, includes the effect of the dielectric material. From

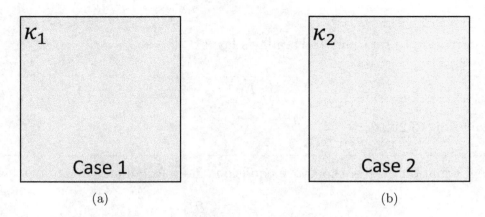

Figure 17.1: A uniform dielectric material of permittivity of the Faraday electric field (a) κ_1, and (b) κ_2.

Equation 17.2, the electric field without the effect of the dielectric material in Case 1 is

$$\vec{D}'_{F1} = \kappa_1 \vec{E}_F, \tag{17.6}$$

and in Case 2,

$$\vec{D}'_{F2} = \kappa_2 \vec{E}_F \neq \vec{D}'_{F1}, \tag{17.7}$$

since $\kappa_1 \neq \kappa_2$ because of the two different dielectric materials assumption.

This result does not make physical sense. The above result means that two different electric fields $\kappa_1 \vec{E}_F$ and $\kappa_2 \vec{E}_F$ are generated by the same source $\vec{B}$, and including the effect of the dielectric material, the same electric field $\vec{E}_F$ is present in both the cases. This is a non-physical result!

In the thought experiment, although Faraday's law and Ampere's law are decoupled, in addition to solving

$$\nabla \times \vec{E}_F = -\frac{\partial \vec{B}}{\partial t}, \tag{17.8}$$

Gauss's law in Equation 13.20, which can be simplified to

$$\nabla \cdot \vec{D}_F = 0, \text{ where } \vec{D}_F = \epsilon_r \epsilon_o \vec{E}_F, \tag{17.9}$$

since $\vec{D}_C$ and ρ are 0, must also be satisfied. Since a uniform isotropic, dielectric material is assumed, ϵ_r can be factored from the $\nabla \cdot$ operator, and the above equation simplifies to

$$\nabla \cdot \vec{E}_F = 0. \tag{17.10}$$

This is the same equation that must be satisfied in both Case 1 and Case 2. Therefore, this does not change the conclusions of the thought experiment results: the same net electric field $\vec{E}_F$ is present in both the cases, and this is non-physical.

This is the motivation to reformulate Faraday's law as

$$\nabla \times \left(\kappa \vec{E}_F\right) = -\frac{\partial \vec{B}}{\partial t}, \tag{17.11}$$

instead of Equation 13.22.

Substituting Equation 17.2 in the above equation, the reformulated Faraday's law is

$$\nabla \times \vec{D}'_F = -\frac{\partial \vec{B}}{\partial t}, \tag{17.12}$$

where

$$\vec{D}'_F = \kappa \vec{E}_F. \tag{17.13}$$

In the reformulated Faraday's law, the electric field without the effect of the dielectric material $\vec{D}'_F$, is now the same in Case 1 and Case 2. From Equation 17.2, the electric field including the effect of the dielectric material in Case 1 is

$$\vec{E}_{F1} = \frac{\vec{D}'_F}{\kappa_1}, \tag{17.14}$$

and in Case 2,

$$\vec{E}_{F2} = \frac{\vec{D}'_F}{\kappa_2} \neq \vec{E}_{F1}. \tag{17.15}$$

Now the results make physical sense. The electric fields $\vec{D}'_F$ generated by the same time-varying magnetic-flux density $\vec{B}$ are equal, but the electric fields $\vec{E}_F$, including the effect of the dielectric materials, are different.

Equation 17.12 is formulated according to Case B, where the electric field $\kappa \vec{E}_F$ generated in the reformulated Faraday's law, does not include the effect of the dielectric material.

Case C is a special case of Case B, where a dielectric material has *no* effect on the electric field generated in Faraday's law. In Equation 17.12, if κ is *exactly* 1.0,

$$\kappa \vec{E}_F = \vec{E}_F, \text{ if } \kappa = 1.0, \tag{17.16}$$

or the electric field generated is the same as the net electric field, including the effect of the dielectric material. In this case, the reformulated Faraday's law reduces to the present formulation in Equation 13.22.

Case C is how the present formulation of Faraday's law can be interpreted: the electric field

generated is *exactly* identical to the net electric field, including the effect of the dielectric material. Case C, however, appears to be too ideal. In the case of the distilled water experiment in Chapter 22, it can be expected that
$$\kappa \approx 1.0, \qquad (17.17)$$
but not precisely equal to 1.0.

18
The Reformulated Maxwell's Equations

Other than Faraday's law, the rest of the electromagnetic equations are unmodified. The set of reformulated Maxwell's equations is

$$\nabla \cdot \left(\vec{D}_C + \vec{D}_F\right) = \rho \tag{18.1}$$

$$\nabla \cdot \vec{B} = 0 \tag{18.2}$$

$$\nabla \times \vec{D}'_F = -\frac{\partial \vec{B}}{\partial t}, \text{ where } \vec{D}'_F = \kappa \vec{E}_F \tag{18.3}$$

$$\nabla \times \vec{E}_C = \vec{0} \tag{18.4}$$

$$\nabla \times \vec{H} = \vec{J} + \frac{\partial}{\partial t}\left(\vec{D}_C + \vec{D}_F\right), \tag{18.5}$$

and the above equations are also valid for time-varying fields, no different from, and for the same reasons as the existing formulation, proven in Reference [3]. More details on the above equations can be found in Chapter 13.

In addition to the above equations,

$$\vec{D}_F = \epsilon_r \epsilon_o \vec{E}_F \tag{18.6}$$

$$\vec{D}'_F = \kappa \vec{E}_F \tag{18.7}$$

$$\vec{D}_C = \epsilon_r \epsilon_o \vec{E}_C \tag{18.8}$$

$$\vec{B} = \mu_r \mu_o \vec{H}, \tag{18.9}$$

capture the effect of a material on the fields. From the permittivity of distilled water example in Chapter 22, it will become clear that the value of ϵ_r in the reformulated equations, may be a little different from the value of ϵ_r in the existing formulation. The permeability μ_r of a material, as discussed in Chapter 19, does not change in the reformulated equations.

The values (and units) of the constants ϵ_o and μ_o are not modified, and are the same as the ones documented in Table 3.2.

18.1 The Law of Conservation of Energy

In Chapter 13, Maxwell's equations were rewritten to separate the electric fields, depending on its source. This has been done to not violate the law of conservation of energy in electrostatics, in the reformulated equations.

There are two equations embedded in Faraday's law,

$$\nabla \times \vec{E} = -\frac{\partial \vec{B}}{\partial t}. \tag{18.10}$$

Writing $\vec{E}$ as the superposition of $\vec{E}_C$ and $\vec{E}_F$ in Equation 13.32,

$$\nabla \times \left(\vec{E}_C + \vec{E}_F\right) = -\frac{\partial \vec{B}}{\partial t}. \tag{18.11}$$

This equation can be separated into two equations, Equation 13.22 and Equation 13.23,

$$\nabla \times \vec{E}_C = \vec{0} \tag{18.12}$$

$$\nabla \times \vec{E}_F = -\frac{\partial \vec{B}}{\partial t}. \tag{18.13}$$

Modifying Faraday's law in Equation 18.13, does not affect Equation 18.12, thereby not violating the law of conservation of energy, explained next.

In electrostatics, where there are no time-varying fields, or current sources, Faraday's law in Equation 18.10 simplifies to

$$\nabla \times \vec{E} = \vec{0}, \tag{18.14}$$

which must be satisfied to not violate the law of conservation of energy in electrostatics, discussed in Section 7.1.

If Equation 18.10 is reformulated as discussed in Chapter 17, without separating the electric field into $\vec{E}_C$ and $\vec{E}_F$,

$$\nabla \times \left(\kappa \vec{E}\right) = -\frac{\partial \vec{B}}{\partial t}. \tag{18.15}$$

In electrostatics, the above equation reduces to

$$\nabla \times \left(\kappa \vec{E}\right) = \vec{0}, \tag{18.16}$$

which is not the same as Equation 18.14, thereby violating the law of conservation of energy.

In the reformulated equations, in electrostatics,

$$\frac{\partial \vec{B}}{\partial t} = \vec{0} \tag{18.17}$$

$$\therefore \vec{E}_F = \vec{0}, \tag{18.18}$$

and Faraday's law is always satisfied, since the left-hand side and the right-hand side of the equation are both $\vec{0}$. Similarly, in Ampere's law,

$$\frac{\partial \vec{D}_C}{\partial t} = \frac{\partial \vec{D}_F}{\partial t} = \vec{0} \tag{18.19}$$

$$\vec{J} = \vec{0} \tag{18.20}$$

$$\therefore \vec{H} = \vec{0}, \tag{18.21}$$

and Ampere's law is also satisfied. The equations to be solved in electrostatics, simplifies to

$$\nabla \cdot \vec{D}_C = \rho \tag{18.22}$$

$$\nabla \times \vec{E}_C = \vec{0}. \tag{18.23}$$

Equation 18.23 enforces the law of conservation of energy. By separating the electric field into $\vec{E}_C$ and $\vec{E}_F$, the reformulation of Faraday's law in Equation 18.3, does not affect Equation 18.4,

$$\nabla \times \vec{E}_C = \vec{0}, \tag{18.24}$$

and the law of conservation of energy in the reformulated equations, in electrostatics, is still met.

19

Consequences of the Reformulation of Faraday's Law

There are both changes and invariance to the electrical characteristics, such as fields, definitions, material properties, etc., as a consequence of modifying Faraday's law. The characteristics that are invariant to the reformulation will be presented first, followed by the changes.

19.1 Characteristics Invariant to the Reformulation of Faraday's Law

Although Faraday's law is reformulated, this does not change the definition of any of the electrical quantities. The flowchart of definitions, explained in detail in Reference [3], is shown in Figure 19.1. Starting with the definition of charge q using Coulomb's law, highlighted in the flowchart, the

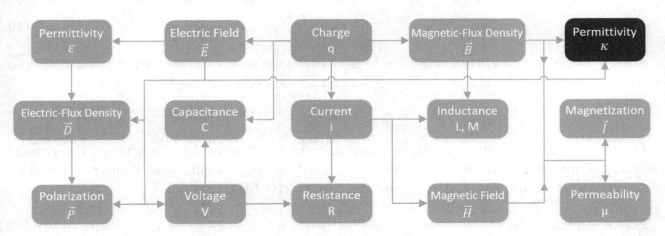

Figure 19.1: The flowchart of definitions in the SI system.

remaining electrical variables can be defined. An outgoing arrow from a variable, means that this variable is used to define the variable that the arrow is incoming to. For example, the definition of current i is

$$i = \frac{\Delta q}{\Delta t}, \tag{19.1}$$

where charge Δq flows during the time interval Δt. In the chart, the incoming arrow to i from q, means that i is defined using charge q, as written in the above equation.

Since the definitions of the electrical quantities are invariant to the reformulation, the unit value of the quantities also remain the same. For example, charge q is defined from Coulomb's law,

$$\vec{F} = \frac{1}{4\pi\epsilon_o} \frac{q_1 q_2}{r^2} \hat{r}, \tag{19.2}$$

where $\vec{F}$ is the force between two point charges q_1 and q_2, separated by distance r, and the force is in the direction of the unit vector $\hat{r}$. Using Coulomb's law, the unit charge of 1 *coulomb* is the quantity, when separated in equal amounts by $1\,m$, repel each other with a force of $1/(4\pi\epsilon_o)\,N$. Similarly, this example is applicable to all the definitions in the flowchart, and the reformulation does not introduce any differences in the unit values.

The Lorentz force law,

$$\vec{F} = q\vec{v} \times \vec{B}, \tag{19.3}$$

where $\vec{F}$ is the force on a charge q, traveling at velocity $\vec{v}$ in the magnetic-flux density $\vec{B}$, is used as the definition of $\vec{B}$. This definition also remains the same in the new formulation. In Reference [3], this equation was derived from Faraday's law. The Lorentz force law can also be derived from the reformulated Faraday's law, presented in Chapter 23.

Alternately, in the flowchart of definitions, current i can be defined first using the equation,

$$\frac{F}{L} = \frac{\mu_o}{2\pi} \frac{i_1 i_2}{d}, \tag{19.4}$$

where F is the force between two wires of negligible diameter, carrying currents i_1 and i_2, and of length L, separated by d [3]. The charge q can then be defined using Equation 19.1,

$$\Delta q = i \Delta t. \tag{19.5}$$

The unit charge is the charge contained in a unit current, flowing for a unit time. The remaining electrical quantities can be defined, following the flowchart of definitions, similar to the electro-magnetic system of units (EMU), where the current is defined first [3].

The new addition to the definitions flowchart is κ, the permittivity of the Faraday electric field $\vec{E}_F$. κ is defined from the reformulated Faraday's law,

$$\nabla \times \left(\kappa \vec{E}_F\right) = -\frac{\partial \vec{B}}{\partial t}. \tag{19.6}$$

The physical meaning of $\vec{D}'_F$,

$$\vec{D}'_F = \kappa \vec{E}_F, \tag{19.7}$$

as discussed in Chapter 17, is the electric field at a point, without the effect of the dielectric material on $\vec{E}_F$ at that point.

No change of units is required in the reformulated equations. Since only Faraday's law is modified, and by definition, κ has no units, the units of all the quantities remain the same as the existing SI units.

The values and units of the constants ϵ_o and μ_o are documented in Table 3.2. These are not modified either, and the reason for picking their values has been discussed in detail in Reference [3].

In Reference [3], and discussed briefly in Chapter 9, the ballistic-galvanometer experiment can be used to characterize the relative permeability μ_r of a material. In this experiment, a quick change in the current in the primary coil Δi, induces a transient current in the secondary coil, which can be detected from the throw of the needle of the ballistic galvanometer in the secondary coil. Δi in the primary coil, results in a change in the magnetic flux in the secondary coil, caused by ΔH in the case of a non-magnetic material, and ΔB in the case of a magnetic material. Sweeping Δi over a range in small steps, the $B - H$ plot can be obtained.

The equation

$$\Delta B = \Delta H \left(\frac{\theta'}{\theta}\right), \qquad (19.8)$$

has been derived in Reference [3], where θ' and θ are the throws of the needle of the ballistic galvanometer, in the case of a magnetic and a non-magnetic toroid. This equation is independent of κ in the reformulated Faraday's law. Therefore, the permeability μ_r that has been characterized of a material, is invariant to the modification of Faraday's law.

19.2 Changes Resulting From the Reformulation of Faraday's Law

Of course, the permittivity of the Faraday electric field κ, is a new parameter introduced. The technique to determine κ, and its value for the example of distilled water will be presented in Chapter 22. It will be shown that there may be a small difference in the permittivity of the Coulomb electric field ϵ_r, between the existing formulation and the reformulated equations.

Since Faraday's law is reformulated, the wave equation will be different, derived in Chapter 20. A small change in the field values can also be expected, depending on the value of κ. Note that in the case where κ is *exactly* equal to 1.0, the equations reduce to the existing formulation. The reformulated Maxwell's equations is a little general formulation of the existing equations.

19.3 The Reformulated Faraday's Law and Magnetic Current Density

Equation 19.7 can be written as a vector sum, instead of a scaling factor κ,

$$\vec{D}'_F = \vec{E}_F + \vec{P}'_F, \qquad (19.9)$$

similar to Equation 14.6, where the relation between the electric-displacement field $\vec{D}_C^*$ and the electric field $\vec{E}_C$ is written using the polarization vector $\vec{P}_C^*$, instead of scaling the electric field by ϵ_r in Equation 14.1.

If the polarization vector $\vec{P}_F'$ is written as

$$\vec{P}_F' = \chi_e' \vec{E}_F, \tag{19.10}$$

Equation 19.9 can be written as

$$\vec{D}_F' = (1 + \chi_e') \vec{E}_F. \tag{19.11}$$

If

$$\kappa = 1 + \chi_e', \tag{19.12}$$

Equation 19.11 is the same as Equation 19.7. χ_e' is similar to the electric susceptibility χ_e in Equation 6.6.

Substituting Equation 19.9 in the reformulated Faraday's law in Equation 17.12, and rearranging the terms,

$$\nabla \times \vec{E}_F = -\nabla \times \vec{P}_F' - \frac{\partial \vec{B}}{\partial t}. \tag{19.13}$$

In Reference [3], the magnetic current density $\vec{K}$ is introduced in Faraday's law, to bridge the asymmetry between Faraday's law and Ampere's law. Although this derivation is presented in CGS units, similar steps can be followed in SI units to derive

$$\nabla \times \vec{E} = -\vec{K} - \frac{\partial \vec{B}}{\partial t}. \tag{19.14}$$

Equating the above equation to Equation 19.13,

$$\vec{K} = \nabla \times \vec{P}_F', \tag{19.15}$$

can be viewed as a magnetic current density, resulting in the generation of the electric field $\vec{E}_F$.

20

The Wave Equation in the Reformulated Maxwell's Equations

In a source-free region, and a uniform medium with permittivity ϵ_r, κ, and permeability μ_r, assuming that $\vec{D}_C$ is 0, Faraday's law and Ampere's law in Equation 18.3 and Equation 18.5, can be simplified to

$$\nabla \times \left(\kappa \vec{E}_F\right) = -\mu_r \mu_o \frac{\partial \vec{H}}{\partial t} \tag{20.1}$$

$$\nabla \times \vec{H} = \epsilon_r \epsilon_o \frac{\partial \vec{E}_F}{\partial t}. \tag{20.2}$$

Since $\vec{D}_C$ is assumed to be 0, and in a source-free region, Gauss's law in Equation 18.1 can be simplified to

$$\nabla \cdot \left(\epsilon_r \vec{E}_F\right) = 0. \tag{20.3}$$

If a uniform material is assumed, ϵ_r can be factored from the $\nabla \cdot$ operator in the above equation, simplifying the equation to

$$\nabla \cdot \vec{E}_F = 0, \tag{20.4}$$

and similarly, factoring κ from the $\nabla \times$ operator in Equation 20.1,

$$\nabla \times \vec{E}_F = -\frac{\mu_r \mu_o}{\kappa} \frac{\partial \vec{H}}{\partial t}. \tag{20.5}$$

Using the identity,

$$\nabla \times \nabla \times \vec{E}_F = \nabla \left(\nabla \cdot \vec{E}_F\right) - \nabla^2 \vec{E}_F, \tag{20.6}$$

and simplifying the above equation using Equation 20.4,

$$\nabla \times \nabla \times \vec{E}_F = -\nabla^2 \vec{E}_F. \tag{20.7}$$

Applying the $\nabla \times$ operation on both sides of Equation 20.5,

$$\nabla \times \nabla \times \vec{E}_F = -\frac{\mu_r \mu_o}{\kappa} \frac{\partial}{\partial t} \left(\nabla \times \vec{H}\right). \tag{20.8}$$

Substituting Equation 20.2 in the above equation,

$$\nabla \times \nabla \times \vec{E}_F = -\frac{\mu_r \epsilon_r \mu_o \epsilon_o}{\kappa} \frac{\partial^2 \vec{E}_F}{\partial t^2}. \tag{20.9}$$

Simplifying the above equation using Equation 20.7,

$$\nabla^2 \vec{E}_F = \frac{\mu_r \epsilon_r \mu_o \epsilon_o}{\kappa} \frac{\partial^2 \vec{E}_F}{\partial t^2}. \tag{20.10}$$

Since

$$\frac{1}{\mu_o \epsilon_o} = c^2 \approx (3.0 \times 10^8)^2 \; m^2/s^2, \tag{20.11}$$

the speed of light squared [3], rewriting Equation 20.10,

$$\nabla^2 \vec{E}_F = \frac{1}{\frac{\kappa c^2}{\mu_r \epsilon_r}} \frac{\partial^2 \vec{E}_F}{\partial t^2}. \tag{20.12}$$

This is the famous wave equation. The solution to this equation are electromagnetic waves at frequency f and wavelength λ, which propagate at a velocity,

$$v = f\lambda$$
$$= \sqrt{\frac{\kappa c^2}{\mu_r \epsilon_r}}$$
$$= \frac{c}{\sqrt{\mu_r \left(\frac{\epsilon_r}{\kappa}\right)}}. \tag{20.13}$$

This is a little different result, compared to the wave equation in the existing formulation. Using similar steps in this chapter, it is left as an exercise for the reader to derive the wave equation using the present formulation of Maxwell's equations,

$$\nabla^2 \vec{E}_F = \frac{1}{\frac{c^2}{\mu_r \epsilon_r}} \frac{\partial^2 \vec{E}_F}{\partial t^2}, \tag{20.14}$$

where the electromagnetic waves propagate with velocity

$$v = f\lambda$$
$$= \sqrt{\frac{c^2}{\mu_r \epsilon_r}}$$
$$= \frac{c}{\sqrt{\mu_r \epsilon_r}}. \tag{20.15}$$

Note that if $\kappa = 1.0$, Equation 20.12 reduces to Equation 20.14.

21

$\kappa \neq \epsilon_r$

In Faraday's experiment with spherical capacitors from Chapter 5, the effect of a dielectric material on the electric field due to electric charges $\vec{E}_C$, is captured in the parameter ϵ_r. The experiment results clearly show that the electric field $\vec{E}_C$ in the dielectric material is the reduced electric field by a factor of ϵ_r, or in other words,

$$\vec{D}_C^* = \epsilon_r \vec{E}_C, \tag{21.1}$$

where $\vec{D}_C^*$ is the electric field at a point, without the effect of the dielectric material at that point. This has also been explained in Chapter 15.

The above interpretation of $\vec{D}_C^*$, as the electric field without the effect of the dielectric material, however, is not applicable to $\vec{D}_F^*$,

$$\vec{D}_F^* = \epsilon_r \vec{E}_F. \tag{21.2}$$

In the case of $\vec{E}_F$, a new equation similar to Equation 21.1, and a new permittivity κ was introduced in Chapter 17,

$$\vec{D}_F' = \kappa \vec{E}_F, \tag{21.3}$$

where $\vec{D}_F'$ is the electric field at a point, without the effect of the dielectric material at that point. By definition, the net electric field at any point, including the effect of the dielectric material at that point, is $\vec{E}_F$.

κ is different from ϵ_r,

$$\kappa \neq \epsilon_r. \tag{21.4}$$

If κ is the same as ϵ_r, ϵ_r and κ cancel each other in Equation 20.12, and the velocity of the electromagnetic wave is independent of the dielectric material. This is clearly an incorrect result, and experimental observations such as Snell's law [13], can no longer be derived from Maxwell's equations.

Clearly, ϵ_r is different from κ. Why is this the case? This question will be answered in the

remainder of this chapter. In the example of the permittivity values of distilled water in Chapter 22,
$$\epsilon_r \approx 80, \qquad (21.5)$$
while
$$\kappa \approx 1. \qquad (21.6)$$
The difference in the permittivity values, can be explained by simple atomic representations of the dielectric material, when immersed in the field patterns $\vec{E}_C$ and $\vec{E}_F$, similar to Chapter 6.

A parallel-plate capacitor is used to illustrate the electric-field pattern due to electric charges. However, the same explanation is also applicable to a spherical capacitor. Two parallel plates, charged positive and negative, sandwiched between a dielectric material, is shown in Figure 21.1(a). The electric field between the charged plates is shown by the arrows.

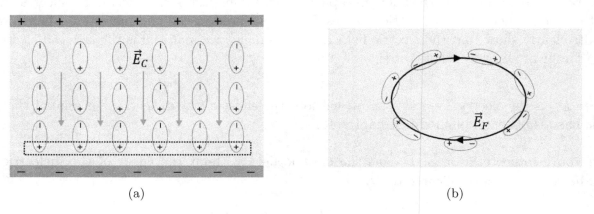

Figure 21.1: The polarization of the atoms in a dielectric material due to the (a) electric field from electric charges, (b) electric field generated by a time-varying magnetic-flux density $\vec{B}$ in Faraday's law.

A simplistic view of the atoms in the dielectric is shown in the figure. The electric field in the dielectric due to the electric charges in the parallel plates, exerts a force on the positive/negative charges in the atoms, distorting them to resemble electric dipoles. This polarization of the atoms in the dielectric material, results in a reduced electric field by a factor of ϵ_r.

Lets compare the electric-field pattern in the parallel-plate capacitor, to the electric-field pattern generated in Faraday's law. The reformulated Faraday's law in Equation 18.3, can be cast in the integral form using the Kelvin-Stokes theorem, similar to the steps in Section 7.1 or Reference [3],
$$\oint_\ell \kappa \vec{E}_F \cdot \vec{dl} = -\frac{\partial}{\partial t} \int_S \vec{B} \cdot d\vec{A}, \qquad (21.7)$$
where S is the area enclosing the loop ℓ. The direction of the path integral, and the direction of the area element vector $d\vec{A}$, are related by the right-hand rule: if the fingers of the right hand, curl in the direction of the path integral, the thumb points in the direction of the vector $d\vec{A}$.

If the electric field $\vec{E}_F$ is "swirling", as shown in Figure 21.1(b), and if the path of integration ℓ is the same as the oval loop shown in the figure, then $\vec{E}_F$ is parallel to ℓ, and therefore,

$$\oint_\ell \kappa \vec{E}_F \cdot \vec{dl} \neq 0. \tag{21.8}$$

Such a "swirling" electric-field pattern can be expected to be generated by a time-varying magnetic-flux density $\vec{B}$ in Faraday's law, since Equation 21.8 holds true in the case of Faraday's law in Equation 21.7.

Note that the net electric field $\vec{E}_C$ in the case of a parallel-plate capacitor, must satisfy Equation 7.2,

$$\oint_\ell \vec{E}_C \cdot \vec{dl} = 0, \tag{21.9}$$

where ℓ is the path of integration of any closed loop, to not violate the law of conservation of energy. It is interesting that the above equation must always be satisfied at any time instant, regardless of whether $\vec{E}_C$ is a static or a time-varying field. The differential form of the above equation in Equation 7.6, is one of the Maxwell's equations in Equation 18.4, which must always hold true at any time instant. Equation 21.9 is the opposite of Equation 21.8, resulting in different electric-field patterns of $\vec{E}_C$ and $\vec{E}_F$.

Similar electric dipoles can be expected in the case of the electric field generated in Faraday's law, shown in the figure. These dipoles are a consequence of the force on the positive and the negative charges making up the atoms of the dielectric material. The orientation of the dipoles, however, are very different compared to Figure 21.1(a). In the simplified representation, there is a net "row" of positive charges shown by the dotted box in Figure 21.1(a), and a periodic arrangement of dipoles in the case of $\vec{E}_C$. This is not the case, however, with $\vec{E}_F$ in Figure 21.1(b).

This difference in the orientation of the electric dipoles, can be used to explain the difference in the effect of the dielectric material on an electric field in modifying the field, and the large difference in the permittivity values between ϵ_r and κ.

21.1 A Note on Ampere's Law

With the introduction of $\vec{D}'_F$, Ampere's law is not reformulated, for example, such as

$$\nabla \times \vec{H} = \vec{J} + \epsilon_o \frac{\partial \vec{D}_C^*}{\partial t} + \epsilon_o \frac{\partial \vec{D}'_F}{\partial t}. \tag{21.10}$$

The problem with this formulation is that the wave equation, following similar steps in Chapter 20, would become independent of the permittivity ϵ_r. As explained earlier in the chapter, this result is incorrect.

From Equation 21.6 and Equation 21.3, the above equation can be written as

$$\nabla \times \vec{H} \approx \vec{J} + \epsilon_o \frac{\partial \vec{D}_C^*}{\partial t} + \epsilon_o \frac{\partial \vec{E}_F}{\partial t}. \tag{21.11}$$

In comparison to the existing formulation of Ampere's law in Equation 14.13,

$$\nabla \times \vec{H} = \vec{J} + \epsilon_o \frac{\partial}{\partial t}\left(\vec{E}_C + \vec{P}_C^*\right) + \epsilon_o \frac{\partial}{\partial t}\left(\vec{E}_F + \vec{P}_F^*\right), \tag{21.12}$$

Equation 21.11 does not include the contribution of the polarization current density $\frac{\partial \vec{P}_F^*}{\partial t}$. From the analysis earlier in the chapter, there is little effect in modifying the electric field in Faraday's law by the dielectric material. Although this may be the case, the existing formulation of Ampere's law, does include the polarization current density $\frac{\partial \vec{P}_F^*}{\partial t}$, which captures the oscillation of the charges in the electric dipoles due to the time-varying $\vec{E}_F$, in the generation of the magnetic field.

21.2 The Dependence of ϵ_r on the Electric-Field Pattern

The permittivity of the Coulomb electric field ϵ_r, from Equation 5.8, is a measure of the modification of an electric field due to electric charges, by the dielectric material. The analysis presented in this chapter shows that there is a dependence between the electric-field pattern, and the effect of the dielectric material in modifying the electric field. A mild dependence, therefore, between the electric-field pattern due to electric charges, and the permittivity ϵ_r measured, can be expected.

For example, in the case of a spherical capacitor, the electric-field pattern is radial, shown in Figure 6.2, as opposed to a vertical electric-field pattern in a parallel-plate capacitor. It would not be surprising to measure a small difference in the permittivity of the dielectric material between these two cases, since the electric-field patterns are different.

22

Measurement of ε_r and κ

A measurement technique to determine the permittivity values ε_r and κ, will be presented in this chapter. In this chapter, the scripted character ε_r is used to denote the permittivity calculated with the reformulated equations in Chapter 18, to differentiate from ϵ_r in the existing formulation in Chapter 13. The measurement data published in References [14]–[15], will be used to calculate ε_r and κ of distilled water.

22.1 Two Ways to Measure ϵ_r in the Existing Formulation of Maxwell's Equations

The different techniques to measure permittivity ϵ_r have been discussed in Reference [16]. By sweeping the frequency and the temperature of the dielectric, the effect of these parameters on permittivity can be characterized.

Two ways to measure permittivity ϵ_r are the following:

1. Measurement of ϵ_r from the ratio of capacitances. This result will be referred to as $\epsilon_{r,Method\,1}$.

2. Measurement of ϵ_r from the electromagnetic wave. This result will be referred to as $\epsilon_{r,Method\,2}$.

In the existing formulation, both these results are assumed to be identical,

$$\epsilon_{r,Method\,1} = \epsilon_{r,Method\,2} = \epsilon_r. \tag{22.1}$$

Any difference in the above two results cannot be explained in the existing formulation. Each of the above two methods are explained next.

22.1.1 Method 1: Ratio of Capacitances

Faraday's experiments with spherical capacitors in Chapter 5, led to the introduction of ϵ_r in Gauss's law. Using Gauss's law, the capacitance of a spherical capacitor or a parallel-plate capacitor, for example, can be calculated [3]. Although the derivation in Reference [3] is in electrostatic units, repeating the same steps using the SI version of Gauss's law in Equation 5.14, the capacitance of a parallel-plate capacitor in Figure 21.1(a) is

$$C_d = \epsilon_r \epsilon_o \frac{A}{d}, \qquad (22.2)$$

where ϵ_r is the permittivity of the dielectric material sandwiched between the plates, A is the surface area of the plate, and d is the distance separating the plates.

The capacitances of two capacitors, one filled with the dielectric material whose ϵ_r is to be characterized, and the second capacitor with air as the dielectric material, are accurately measured. In the case of air as the dielectric material, Equation 22.2 simplifies to

$$C_a = \epsilon_o \frac{A}{d}. \qquad (22.3)$$

Strictly speaking $\epsilon_r = 1.0$ for vacuum, but air is used for simplicity. The ratio of the capacitances,

$$\epsilon_r = \frac{C_d}{C_a}, \qquad (22.4)$$

is the permittivity ϵ_r of the material to be characterized. It is assumed that the entire electric field of the capacitor is present in the dielectric material. Otherwise, Equation 22.4 needs to be modified to take this into account. The details of the experiment are provided in Reference [14].

22.1.2 Method 2: Electromagnetic Wave

An alternate method to determine the value of ϵ_r is from electromagnetic waves. A non-magnetic material, where $\mu_r = 1$, will be assumed for simplicity. If the wavelength of an electromagnetic wave λ_d in a dielectric material, whose permittivity ϵ_r is to be measured, has been determined, from Equation 20.15,

$$\lambda_d = \frac{c}{f\sqrt{\epsilon_r}}. \qquad (22.5)$$

Repeating the same experiment in the dielectric medium air (or vacuum to be precise), where $\epsilon_r = 1$, and at the same frequency, the wavelength of the electromagnetic wave is

$$\lambda_a = \frac{c}{f}. \qquad (22.6)$$

From the above equations, ϵ_r can be calculated from the ratio of the wavelengths,

$$\epsilon_r = \left[\frac{\lambda_a}{\lambda_d}\right]^2. \qquad (22.7)$$

Such a scheme has been used in Reference [15], to measure the permittivity of distilled water from electromagnetic waves in a coaxial transmission line. Although transmission lines are beyond the scope of this book, only the results from Reference [15] will be presented in Section 22.3. Additional information on this paper can be found in Reference [17].

In the case of a lossless coaxial transmission line, the propagation velocity in Equation 20.15 can be derived [18]. Including the loss in a transmission line, which are the non-zero resistivity of the conductors and the dielectric material, Equation 20.15, and consequently Equation 22.7, need to be modified to take the loss into account [18].

The experiments in References [14]–[15] can be repeated at different frequencies, to show that the permittivity ϵ_r varies as a function of frequency.

22.2 Measurement of ε_r and κ in the Reformulated Maxwell's Equations

The permittivity values ε_r and κ in the reformulated equations, can be determined from the two experiments labeled *Method* 1 and *Method* 2 in Section 22.1. Unlike the existing formulation, which assumes that the results from the two methods are identical, in the reformulated equations, *Method* 1 and *Method* 2 are successively applied to yield ε_r and κ.

Method 1 can be used to determine ε_r. The capacitance equations, Equation 22.2 and Equation 22.3, are the same in the reformulated equations. From the ratio of the measured capacitances, the permittivity ε_r can be calculated,

$$\varepsilon_r = \frac{C_d}{C_a}, \tag{22.8}$$

which is the same as Equation 22.4.

Method 2 can be used to determine κ, after ε_r has been determined using *Method* 1. The wavelengths in a dielectric material λ_d, and in air λ_a, can be measured, similar to Reference [15]. These values do not depend on whether the existing formulation of Maxwell's equations, or the reformulated equations are used. As discussed in Chapter 19, the unit current, the unit voltage, the unit $\vec{B}$, etc. are invariant to the modification of Faraday's law.

When the dielectric material is air, Equation 20.13 reduces to Equation 22.6,

$$\lambda_a = \frac{c}{f}, \tag{22.9}$$

since ϵ_r, κ, and μ_r are equal to 1.0. From Equation 20.13, in the case of a non-magnetic material,

$$\lambda_d = \frac{c}{f\sqrt{\frac{\varepsilon_r}{\kappa}}}. \tag{22.10}$$

From the above equations,

$$\frac{\varepsilon_r}{\kappa} = \left[\frac{\lambda_a}{\lambda_d}\right]^2. \tag{22.11}$$

Since ε_r is known from *Method 1*, this value can be used to calculate κ from the above equation,

$$\kappa = \varepsilon_r \left[\frac{\lambda_d}{\lambda_a}\right]^2. \tag{22.12}$$

The comparison of Equation 22.7 and Equation 22.11 shows that the measurement of permittivity ϵ_r in the existing formulation using electromagnetic waves, is the same as the measurement of ε_r/κ in the reformulated Maxwell's equations. Since ε_r is a function of frequency, as noted in Section 22.1.2, it can be expected that κ is a function of frequency as well.

As discussed in Section 22.1.2, Equation 22.11 holds true only in the case of a lossless transmission line. In the case of the lossy transmission line in Reference [15], this equation will be used as an approximation to estimate the value of κ in distilled water, in the next section.

22.3 ε_r and κ of Distilled Water

The goal of the exercise is not to calculate the most accurate value of κ, but an estimate of its value for distilled water. Several approximations will be made in determining ε_r and κ of distilled water:

1. The measurement results of water, documented in Table 22.1, will be used in the calculations. The same water may not have been used in both the experiments.
2. Although Equation 22.12 is only valid in the case of a lossless transmission line, this equation will be used to estimate the value of κ in a lossy transmission line in Reference [15].
3. The frequency of measurement in Reference [14] is not the same as Reference [15]. This will also introduce an error in the calculations.

Source	Permittivity	Value	Frequency
Ref. [14]	$\epsilon_{r,Method\,1}$	78.3	3 – 96 kHz
Ref. [15]	$\epsilon_{r,Method\,2}$	78.57	12 – 77 MHz

Table 22.1: The permittivity values used in the calculation of κ in distilled water.

As discussed in Section 22.2,

$$\epsilon_{r,Method\,1} \Rightarrow \varepsilon_r, \tag{22.13}$$

the measurement of $\epsilon_{r,Method\,1}$ in the existing formulation, is the same as the value of ε_r in the reformulated equations. Therefore, from the data in the first row in Table 22.1,

$$\varepsilon_r = 78.3. \tag{22.14}$$

As discussed in Section 22.2,

$$\epsilon_{r,Method\,2} \Rightarrow \frac{\varepsilon_r}{\kappa}, \tag{22.15}$$

the measurement of $\epsilon_{r,Method2}$ in the existing formulation, is the same as the value of ε_r/κ in the reformulated equations. Using the data in the second row of Table 22.1, and using the value of ε_r in Equation 22.14,

$$\kappa = 0.9966 \neq 1.0. \tag{22.16}$$

Many approximations have been made in the calculation of κ, and with the uncertainty in the measurement values, more experiments are needed to verify the reformulation of Faraday's law.

Note that the estimated value of κ is less than 1,

$$\kappa < 1. \tag{22.17}$$

From Equation 17.2, this means that the electric field without the effect of the dielectric material $\vec{D}'_F$, is less than the net electric field $\vec{E}_F$, including the effect of the dielectric material,

$$\left|\vec{D}'_F\right| < \left|\vec{E}_F\right|. \tag{22.18}$$

In other words, the dielectric material increases the strength of $\vec{D}'_F$, the electric field generated from a time-varying magnetic-flux density $\vec{B}$. This behavior is similar to a magnetic-field ($\vec{H}$) strength increasing in a magnetic material, resulting in $\vec{B}$, as seen in the ballistic-galvanometer experiment in Chapter 9. This is unlike, however, the effect of a dielectric material on the electric field due to electric charges, which reduces the strength of the electric field, as seen in Equation 5.8. This is reflected in the value of ϵ_r,

$$\epsilon_r \geq 1. \tag{22.19}$$

22.4 Experimental Verification of the Reformulated Maxwell's Equations

Any difference between the values of $\epsilon_{r,Method1}$ and $\epsilon_{r,Method2}$, cannot be explained in the existing formulation, and serves as a proof of the reformulated Faraday's law.

> It is easy to dismiss a tiny difference between the values of $\epsilon_{r,Method1}$ and $\epsilon_{r,Method2}$ as an error. However, there is perhaps, a deep physical meaning underlying the inconsistency between the two values.

In this experiment, there may be an added complexity from the dependence of the permittivity $\epsilon_{r,Method1}$ on the electric-field pattern, discussed in Section 21.2. The measurement results of $\epsilon_{r,Method1}$, repeated on a parallel plate and a spherical capacitor, can be used to understand the difference in their values, if any.

23

The Derivation of the Lorentz Force Law from the Reformulated Faraday's Law

The Lorentz force law, the force on a charge q, moving at velocity $\vec{v}$ in the magnetic-flux density $\vec{B}$ field, is

$$\vec{F} = q\vec{v} \times \vec{B}. \tag{23.1}$$

This equation can be derived from Faraday's law, presented in detail in Reference [3], and is valid at any time instant when $\vec{B}$ is a time-varying field. The Lorentz force law serves as the definition of magnetic-flux density $\vec{B}$.

With the modified Faraday's law,

$$\nabla \times \vec{D}'_F = -\frac{\partial \vec{B}}{\partial t}, \text{ where } \vec{D}'_F = \kappa \vec{E}_F, \tag{23.2}$$

the Lorentz force can still be derived. This derivation will be presented next. Therefore, the definition of $\vec{B}$ is invariant to the reformulation of Faraday's law.

Instead of repeating the derivation in Reference [3] from the beginning, the proof will continue from a key conclusion. It has been shown in the reference, any object, such as a wire segment or a point charge, moving at velocity $\vec{v}$ in the field $\vec{B}$ (static or time varying), at any time instant, experiences the effect of an electric field

$$\vec{E}_F = \vec{v} \times \vec{B}. \tag{23.3}$$

The above equation has been derived from the existing formulation of Faraday's law. In the case of the reformulated Faraday's law, following the exact same steps in the derivation, the above equation is modified as

$$\vec{D}'_F = \vec{v} \times \vec{B}, \text{ where } \vec{D}'_F = \kappa \vec{E}_F. \tag{23.4}$$

Since κ has no units, $\vec{D}'_F$ has the same units as the electric field. $\vec{D}'_F$ can be viewed as an electric field: the electric field at a point, without the effect of the dielectric material at that point, as discussed earlier in Chapter 17. The conclusion in the reference still holds true: any object, such as a wire segment or a point charge, moving at velocity $\vec{v}$ in the field $\vec{B}$, experiences the effect of an electric field, which is $\vec{D}'_F$ in the reformulated equations.

By definition, the force on a unit point charge q in the electric field $\vec{D}'_F$, from Equation 5.2, is
$$\vec{F} = q\vec{D}'_F. \tag{23.5}$$
Substituting the above equation in Equation 23.4, the Lorentz force law
$$\vec{F} = q\vec{v} \times \vec{B}, \tag{23.6}$$
can be derived from the reformulated Faraday's law.

24

Key Points on the Differences Between the Electric-Displacement Fields $\vec{D}_C$ and $\vec{D}_F$

In the case of the electric displacement $\vec{D}_C^*$ and the electric field $\vec{E}_C$ due to electric charges, repeating Equation 14.1, there is only one equation relating them,

$$\vec{D}_C^* = \epsilon_r \vec{E}_C. \tag{24.1}$$

This equation can be understood with the following meanings:

1. As discussed in Chapter 15, $\vec{D}_C^*$ is the electric field without the effect of the dielectric material on $\vec{E}_C$. $\vec{E}_C$ is the net electric field at a point, including the effect of the dielectric material at that point.

2. Alternately, this equation can also be understood from the standpoint of the displacement current in Ampere's law. Repeating Equation 14.15,

$$\vec{J}_C = \epsilon_o \frac{\partial \vec{D}_C^*}{\partial t} \tag{24.2}$$

$$= \epsilon_o \frac{\partial}{\partial t} \left(\vec{E}_C + \vec{P}_C^* \right), \tag{24.3}$$

Equation 24.1 accounts for the displacement current density due to the electric field, and the polarization of the dielectric material, discussed in Chapter 8.

The permittivity ϵ_r is applicable to both the above physical meanings associated with Equation 24.1.

There are two equations, however, relating the electric field $\vec{E}_F$ to the electric-displacement field, unlike $\vec{E}_C$, where there is only one equation, Equation 24.1. The first equation, repeating Equation 17.2, is

$$\vec{D}_F' = \kappa \vec{E}_F. \tag{24.4}$$

Repeating Equation 18.3, κ can be mathematically defined as the permittivity that satisfies

$$\nabla \times \left(\kappa \vec{E}_F\right) = -\frac{\partial \vec{B}}{\partial t}. \tag{24.5}$$

The physical meaning associated with Equation 24.4, discussed in Chapter 17, is that $\vec{D}'_F$ is the electric field at any point, without the effect of the dielectric material at that point. The net electric field $\vec{E}_F$ at a point, includes the effect of the dielectric material at that point.

The second equation, repeating Equation 14.2, is

$$\vec{D}^*_F = \epsilon_r \vec{E}_F. \tag{24.6}$$

This equation can be understood from the standpoint of the displacement current in Ampere's law. Repeating Equation 14.16,

$$\vec{J}_F = \epsilon_o \frac{\partial \vec{D}^*_F}{\partial t} \tag{24.7}$$

$$= \epsilon_o \frac{\partial}{\partial t}\left(\vec{E}_F + \vec{P}^*_F\right), \tag{24.8}$$

Equation 24.6 accounts for the displacement current due to the electric field $\frac{\partial \vec{E}_F}{\partial t}$, and the bound current due to the polarization of the dielectric material $\frac{\partial \vec{P}^*_F}{\partial t}$, discussed in Chapter 8.

24.1 The Total Electric-Displacement Field $\vec{D}^*$ and $\vec{D}'$

From the above discussion, a new equation of the total electric-displacement field, can be formulated. The total electric displacement at any point $\vec{D}^*$, accounting for both $\vec{D}^*_C$ and $\vec{D}'_F$,

$$\vec{D}' = \vec{D}^*_C + \vec{D}'_F \tag{24.9}$$

$$= \epsilon_r \vec{E}_C + \kappa \vec{E}_F \tag{24.10}$$

$$= \vec{E}_{air}, \tag{24.11}$$

is the same as $\vec{E}_{air}$, which is the total electric field without the effect of the dielectric material at that point.

The total electric-displacement field $\vec{D}^*$,

$$\vec{D}^* = \vec{D}^*_C + \vec{D}^*_F \tag{24.12}$$

$$= \epsilon_r \vec{E}_C + \epsilon_r \vec{E}_F, \tag{24.13}$$

cannot be interpreted in the same manner as Equation 24.11. $\vec{D}^*$ in the above equation, is present in Ampere's law in Equation 14.12, and accounts for the total displacement current density, the sum of $\vec{J}_C$ and $\vec{J}_F$ in the above equations.

Alternately, from a mathematical perspective, $\vec{D}^*$ in Gauss's law in Equation 14.18, can be viewed as the field that satisfies the equation,

$$\nabla \cdot \left(\epsilon_o \vec{D}^*\right) = \rho. \tag{24.14}$$

25

A Comparison of the $\vec{B}/\vec{H}$ and the $\vec{D}/\vec{E}$ Relations

To summarize the discussion until now, there are two different types of permittivity, the permittivity of the Coulomb electric field ϵ_r, and the permittivity of the Faraday electric field κ, that describe the relation between the electric displacement and the electric field. The reason for introducing two different types of permittivity, discussed in Chapter 21, is a consequence of the differences in the effect of a dielectric material in modifying (1) the electric field due to electric charges, and (2) the electric field due to a time-varying magnetic-flux density $\vec{B}$.

Repeating Equation 14.1,
$$\vec{D}_C^* = \epsilon_r \vec{E}_C, \tag{25.1}$$
describes the relation between the electric displacement $\vec{D}_C^*$ and the electric field $\vec{E}_C$, due to the electric field of electric charges. $\vec{D}_C^*$ is the electric field at a point, without the effect of the dielectric material at that point, discussed in Chapter 15.

Similarly, repeating Equation 17.2,
$$\vec{D}_F' = \kappa \vec{E}_F, \tag{25.2}$$
describes the relation between the electric displacement $\vec{D}_F'$, and the electric field $\vec{E}_F$, due to the electric field generated by a time-varying magnetic-flux density. In this case, $\vec{D}_F'$ is the electric field at a point, without the effect of the dielectric material at that point, discussed in Chapter 17.

However, there is only one type of permeability μ_r, which describes the relation between the magnetic-flux density $\vec{B}$ and the magnetic field $\vec{H}$. This makes the $\vec{B}/\vec{H}$ relation simpler than the $\vec{D}/\vec{E}$ relation. Repeating Equation 14.3,
$$\vec{B}^* = \mu_r \vec{H}. \tag{25.3}$$

There are two sources of the magnetic field $\vec{H}$: an electric current, and a magnet. It was shown in Reference [3], a current loop behaves like a magnet. The magnetic field from the two sources, and therefore, the effect of a magnetic material on the magnetic field, can also be expected to be similar.

The ballistic galvanometer experiment in Chapter 9, clearly demonstrates the effect of a magnetic material on the magnetic field due to a current source. This is captured in Equation 25.3. Since the magnetic field of the two sources are expected to be similar, Equation 25.3 is applicable to both the sources of the magnetic field, requiring only one type of permeability μ_r to capture the effect of a magnetic material on the magnetic field. This is unlike the permittivity, requiring ϵ_r and κ, to capture the effect of a dielectric material on the electric field, depending on the source of the electric field.

26

Other Possibilities in the Reformulation of Maxwell's Equations

If Faraday's law is reformulated as Equation 18.3,

$$\nabla \times \left(\kappa \vec{E}_F\right) = -\frac{\partial \vec{B}}{\partial t}, \tag{26.1}$$

following the same steps in Chapter 20, it can be verified that Ampere's law in Equation 18.5, reformulated as

$$\nabla \times \vec{H} = \vec{J} + \frac{\partial \vec{D}_C}{\partial t} + \kappa \frac{\partial \vec{D}_F}{\partial t}, \tag{26.2}$$

can be used to derive the same wave equation in the present formulation of Maxwell's equations, Equation 20.14,

$$\nabla^2 \vec{E}_F = \frac{1}{\frac{c^2}{\mu_r \epsilon_r}} \frac{\partial^2 \vec{E}_F}{\partial t^2}. \tag{26.3}$$

The current continuity equation,

$$\nabla \cdot \vec{J} = -\frac{\partial \rho_v}{\partial t}, \tag{26.4}$$

where $\vec{J}$ is the volume current density, and ρ_v is the volume charge density, can be derived from Gauss's law and Ampere's law [3]. This serves as a logical proof of Ampere's law and Gauss's law for time-varying fields, discussed in detail in the reference. If Ampere's law is modified as Equation 26.2, Gauss's law in Equation 18.1 must also be modified as

$$\nabla \cdot \left(\vec{D}_C + \kappa \vec{D}_F\right) = \rho_v, \tag{26.5}$$

so that the current continuity equation can be derived. Note that the above reformulated equations of Faraday's law, Ampere's law, and Gauss's law, can be derived from the existing set of Maxwell's equations in Equation 13.20 – Equation 13.24, by replacing

$$\vec{E}_F \to \kappa \vec{E}_F. \tag{26.6}$$

The equations in this chapter are different from the proposed reformulation in Chapter 18. Two different ways to measure permittivity ϵ_r was presented in Chapter 22, labeled $\epsilon_{r,Method1}$ and $\epsilon_{r,Method2}$. The reformulated equations presented in this chapter, cannot explain any difference between the two values of permittivity, and assumes them to be equal. This can be used to prove, the reformulated equations in Chapter 18 is correct.

27

Summary

The answers to the 3 questions at the beginning, in the section "About this Book", can be found in the reformulated Faraday's law Equation 18.3,

$$\nabla \times \left(\kappa \vec{E}_F\right) = -\frac{\partial \vec{B}}{\partial t}. \tag{27.1}$$

1. Why is Faraday's law formulated using the electric field $\vec{E}$, instead of the electric displacement (electric-flux density) $\vec{D}$?

   The existing formulation of Faraday's law using the electric field $\vec{E}$ is a special case, where the dielectric material is assumed to have absolutely no effect on the electric field generated from a time-varying magnetic-flux density $\vec{B}$, discussed in Chapter 17.

2. What effect does a dielectric material have on the electric field generated from Faraday's law? How is this taken into account in Maxwell's equations?

   Equation 17.13 captures the effect of a dielectric material on the electric field from Faraday's law,

   $$\vec{D}'_F = \kappa \vec{E}_F, \tag{27.2}$$

   where the electric displacement $\vec{D}'_F$ at any point, is the electric field without the effect of the dielectric material at that point. $\vec{E}_F$ at any point, is the electric field including the effect of the dielectric material at that point, discussed in Chapter 17. The effect of the dielectric material on the electric field in Faraday's law, is taken into account in the reformulated Maxwell's equations, using the above relation.

3. What is the meaning of $\vec{D} = \epsilon_r \vec{E}$ (written in electrostatic units) in the context of Faraday's law?

   The meaning of this equation can be understood in the context of Ampere's law, where Equation 14.2,

   $$\vec{D}^*_F = \epsilon_r \vec{E}_F, \tag{27.3}$$

   includes the contributions of the polarization current density, and the time-varying electric field $\vec{E}_F$, in the generation of the magnetic field, discussed in Chapter 24. Equation 27.2 is more relevant in the context of Faraday's law, explained in the answer to the second question.

References

[1] Wouter Ö. Koot, *Kepler's Battle with the Mars Orbit. A Modern Approach to the Steps Taken by Kepler*, Utrecht University thesis, Sep. 30, 2014, [Online] Available: https://dspace.library.uu.nl/handle/1874/302355 [Accessed: Sep. 12, 2020].

[2] Johannes Kepler (Author), William H. Donahue (Translator), *Astronomia Nova*, Green Lion Press, New Revised Edition, Dec. 7, 2015.

[3] Krishna Srinivasan, *An Electrifying Introduction to Electromagnetics*, printed by Ingram Spark, Sep. 22, 2018.

[4] Douglas L. Cohen, *Demystifying Electromagnetic Equations: A Complete Explanation of EM Unit Systems and Equation Transformations*, Bellingham, Washington: The Society of Photo-Optical Instrumentation Engineers, 2001.

[5] Sir J. J. Thomson, *Elements of the Mathematical Theory of Electricity and Magnetism*, pp. 120-124, London, UK: Cambridge University Press, 1909.

[6] O. Darrigol, *Electrodynamics from Ampere to Einstein*, pp. 88–90, Oxford U. Press, Oxford, 2000.

[7] Thanu Padmanabhan, *Sleeping Beauties in Theoretical Physics: 26 Surprising Insights*, Springer, 2015.

[8] Ruth W. Chabay and Bruce A. Sherwood, *Matter and Interactions, Volume II: Electric and Magnetic Interactions*, 3rd Edition, Wiley, Jan. 5, 2010.

[9] Richard Fitzpatrick, 'The Vector Product', [Online] Available: http://farside.ph.utexas.edu/teaching/302l/lectures/node6.html [Accessed: July 14, 2020].

[10] 'Triple Product', [Online] Available: https://en.wikipedia.org/wiki/Triple_product [Accessed: July 14, 2020].

[11] Richard Fitzpatrick, 'Gauge Transformations', [Online] Available: http://farside.ph.utexas.edu/teaching/em/lectures/node45.html [Accessed: Apr. 11, 2018].

[12] Evgeny Y. Tsymbal, 'Section 6: Electromagnetic Radiation', [Online] Available: https://unlcms.unl.edu/cas/physics/tsymbal/teaching/EM-914/section6-Electromagnetic_Radiation.pdf [Accessed: Apr. 8, 2018].

[13] Fawwaz T. Ulaby, *Fundamentals of Applied Electromagnetics*, Prentice Hall, Dec. 1998.

[14] C. G. Malmberg and A. A. Maryott, "Dielectric Constant of Water from 0° to 100°C", *Journal of Research of the National Bureau of Standards*, Vol. 56, No. 1, Research Paper 2641, Jan. 1956.

[15] F. H. Drake, G. W. Pierce, and M. T. Dow, "Measurement of the Dielectric Constant and Index of Refraction of Water and Aqueous Solutions of KCL at High Frequencies", *Physical Review*, Vol. 35, pp. 613-622, Mar. 1930.

[16] Xiang Li and Yan Jiang, 'Design of a Cylindrical Cavity Resonator for Measurements of Electrical Properties of Dielectric Materials', Dissertation, 2010, [Online] Available: http://urn.kb.se/resolve?urn=urn:nbn:se:hig:diva-7687 [Accessed: Aug. 3, 2020].

[17] George W. Pierce, "A Table and Method of Computation of Electric Wave Propagation, Transmission Line Phenomena, Optical Refraction, and Inverse Hyperbolic Functions of a Complex Variable", *Proceedings of the American Academy of Arts and Sciences*, Vol. 57, No. 7, pp. 175-191, Apr. 1922. [Online] Available: https://www.jstor.org/stable/pdf/20025901.pdf.

[18] Clayton R. Paul, *Analysis of Multiconductor Transmission Lines*, 2nd Edition, Wiley-IEEE Press, Oct. 26, 2007.

CPSIA information can be obtained
at www.ICGtesting.com
Printed in the USA
LVHW060057041120
670659LV00007B/422